T0318622

Economic Evaluation in
GENOMIC MEDICINE

Economic Evaluation in
GENOMIC MEDICINE

VASILIOS FRAGOULAKIS
University of Patras School of Health Sciences,
Department of Pharmacy, Patras, Greece;
National School of Public Health,
Athens, Greece

CHRISTINA MITROPOULOU
Erasmus University Medical Center,
Department of Clinical Chemistry,
Rotterdam, The Netherlands

MARC S. WILLIAMS
Geisinger Health System, Danville,
Pennsylvania, USA

GEORGE P. PATRINOS
University of Patras School of Health Sciences,
Department of Pharmacy, Patras, Greece

AMSTERDAM • BOSTON • HEIDELBERG • LONDON
NEW YORK • OXFORD • PARIS • SAN DIEGO
SAN FRANCISCO • SINGAPORE • SYDNEY • TOKYO
Academic Press is an imprint of Elsevier

Academic Press is an imprint of Elsevier
32 Jamestown Road, London NW1 7BY, UK
525 B Street, Suite 1800, San Diego, CA 92101-4495, USA
225 Wyman Street, Waltham, MA 02451, USA
The Boulevard, Langford Lane, Kidlington, Oxford OX5 1GB, UK

ISBN: 978-0-12-801497-4

British Library Cataloguing-in-Publication Data
A catalogue record for this book is available from the British Library

Library of Congress Cataloging-in-Publication Data
A catalog record for this book is available from the Library of Congress

For information on all Academic Press publications
visit our website at http://store.elsevier.com/

Typeset by MPS Limited, Chennai, India
www.adi-mps.com

Working together
to grow libraries in
developing countries

www.elsevier.com • www.bookaid.org

Publisher: Christine Minihane
Acquisition Editor: Catherine Van Der Laan
Editorial Project Manager: Lisa Eppich
Production Project Manager: Lucía Pérez
Designer: Maria Inês Cruz

CONTENTS

PREFACE

The subject of this textbook is the emerging scientific field of applied economic evaluation of genomic medicine with a focus on cost-effectiveness analysis, which constitutes the most important part of health economics from a research standpoint. Given that this field is relatively new, we thought that its concepts should be presented in a more intuitive and less formalized manner to familiarize the reader with the field's tools and methods.

Because this is an emerging field bridging two distinct and unrelated disciplines, those of economics and genomic medicine, we have tried to make our text understandable to a broad audience. In particular, this book is addressed to the competent clinical and biomedical scientist, student, or health care professional who possesses a very basic level of knowledge about economic evaluation and would like to delve deeper to be able to evaluate work by others or to apply that knowledge to his or her own studies. This book is also addressed to the economist and economics students who envision acquiring training in health economics with an emphasis on genomic medicine. In this work, we have focused on what we thought was necessary with a thorough presentation of basic concepts and approaches.

The readers of this book should bear in mind that economic evaluation is a highly technical subject that requires a significant investment of time and effort to be fully understood. We believe that this will become apparent as one reads the book. At present, economic evaluation is still at the stage of developing new approaches, whether these are entirely new or borrowed from other related fields such as statistics, mathematics, or econometrics. It is our firm belief that only by understanding its methodologies, including their weaknesses and capabilities, can we make a true and valid judgment of the capabilities of the field itself at the *theoretical* level. Naturally, we should not forget that, despite the attractive veneer of objectivity given by the concise and elegant mathematical nomenclature, the actual subject of financial resource management is *in practice* fundamentally a political problem. Unfortunately, quantitative methods are inadequate for such problems and, in many cases, the translation of "knowledge" into "political decision" involves other factors that are mostly outside the province of the academic community. In this book, however, we have kept the researcher's point of view and have only superficially dealt with certain political issues to contribute impartially, we hope, to the reader's critical thinking.

With regard to the text itself, we offer this suggestion: while reading, do not skip any words, symbols, or gene and biomarker names you do not understand. The reason why most people abandon the study of any subject is because they skip over a word or a symbol they cannot decipher and thus become unable to apply or understand what they are reading.[1] If you find that you are having trouble understanding the text, go back, find a point where you had complete conceptual understanding, and look for a word whose definition you do not know. As soon as you locate it, use a suitable dictionary (either an economics dictionary, if the word is a term of the nomenclature; a biologic or medical dictionary,[2] if the term relates to genetics; or even a conventional dictionary for an "ordinary" word) and clarify it. Both economic evaluation and genomic medicine use a few specialized terms that need to be genuinely understood and not just memorized. You should also keep in mind the fact that evaluating the data presented here is just as important, if not more, than simply learning it. As you study this text, we hope that you will remain critical, that you will remember all that you consider to be true or useful, and that you will compare the information you read with your own personal observations.

We hope that the time you spend reading the text will be worth it, and that it will also give you some ideas for further study or application of the subject. If you achieve this, then we might say your effort had high personal utility and was a cost-effective choice for you and for us as well.

The Authors
January 2015

---

[1] http://www.appliedscholastics.org. Applied Scholastics International, a nonprofit educational organization based in Missouri, was founded by a consortium of American educators in 1972. Administered by the Association for Better Living and Education, it is dedicated to the broad implementation of learning tools researched and developed by American author and educator L. Ron Hubbard.

[2] Genetics Home Reference (http://www.ghr.nlm.nih.gov/glossary) and National Human Genome Research Institute (http://www.genome.gov/glossary).

CHAPTER 1

Economic Evaluation in Health Care: Evidence-Based Medicine and Evidence-Based Health Economics

THE SCOPE OF ECONOMIC EVALUATION OF HEALTH SERVICES: THE SCIENCE OF ECONOMICS

Economic evaluation of health services is a branch of economics that deals with "the systematic evaluation of the benefits and costs arising from the comparison of different health technologies." Economic evaluation is a way of thinking and problem-solving rather than a sterile set of terms or methods used by health economists. Before moving on to its subject matter, we should define its general framework, which is economics itself.

It is a widely held opinion that economics deals exclusively with the processes of production and distribution of wealth, as well as the properties and origin of that well-known means of transaction for a society: money. This approach could be made even broader. The object of economics is primarily the combination of options we should adopt to maximize our welfare under conditions of limited resources. We do not analyze the term "welfare" from an ethical or social point of view, but the term is used liberally by economists to describe the degree of euphoria, exultation, and pleasure produced specifically by the consumption of commodities and services available in the market.

Households, employees, and businesses all face similar "survival" problems and are forced to make certain choices regarding the use of the available resources, regardless of whether the society they live in follows a tender-based or an exchange-based economy. According to economics, the behavior of all these actors should be determined by certain simple tenets, including:

1. That the economic actor wishes to survive and achieve an infinite level of welfare (maximize welfare). This does not include people who do not wish to survive or who do not think rationally. The concept of "welfare" usually considers the person in an "individualistic" sense, with no regard to altruism or social responsibility.
2. In the standard case, economics assumes that "greater consumption of commodities leads to greater welfare," with no consideration of issues of saturation.
3. That the combined consumption of commodities and services (that can be purchased with the available resources) can increase or decrease welfare depending on the ratio, and that there is a specific mix of choices that is ideal for each case (because it maximizes welfare), depending on income and preferences.
4. That resources are limited and economic commodities have a certain cost.
5. That economic stakeholders are willing to sacrifice part of the consumption of a commodity they possess in large quantities to obtain a small quantity of another desirable commodity that they consume at a smaller degree. In economics, this is called "the principle of convexity" because the diagrammatic description of this principle uses a convex mathematical curve.

    Tenet 5 is based on the economics concept that the consumption of any desirable commodity provides welfare that is great at first, but gradually decreases (decreasing utility); therefore, it would be wise for one to refrain from this additional consumption (and the associated cost) to consume something else, which would give them greater welfare. This principle indicates that, generally speaking, economic stakeholders avoid "unilateral consumption" and prefer to consume "a little of each" commodity, but not necessarily in the same ratio.

6. There are specific preferences that can be clearly defined.

From these simple "polished" tenets, we see that achieving infinite welfare is an unobtainable goal, whereas the maximization of welfare with the available means is achievable. For example, a worker follows his or her own maximization pattern and spends any available money to buy a "basket" of commodities (food, clothes, entertainment, health care, etc.), balancing market prices against his or her own preferences.

Without going further into the principles of economics, we should also note that a certain level of welfare can be achieved by different combinations of commodities associated with different costs, and that there is a specific combination that achieves a target level of welfare with minimum cost. For example, if we were interested in an individual's welfare, we could identify several combinations of the commodities "consumption of health care services" and "available income for other services" that would provide a specific target level of welfare. Naturally, we could always achieve this by increasing the consumption of one commodity at the expense of the other, and vice versa, for each combination of choices. To make such an estimate, however, the stakeholders' preferences and the commodities' prices must be expressly specified in each case.

By expanding the application of this example, we can see how a society may decide that it can achieve maximum welfare through a public health system that provides free health care at the expense of "sacrificing" part of the citizens' income, because insurance and taxation costs will take a part of the income available for consumption to invest it in public health institutions. Other societies with more privately oriented health systems provide "less public health care" but compensate for this by taking less income, leaving the rest available for consumption and savings. In accordance with tenet 5, most societies avoid the extreme choices and do not accept a "fully private" or "fully public" character for their health care, instead selecting a mixture of the two. This analysis is simplistic and does not take into account special interest groups or the political scenery; however, it is obvious that each country, financial organization, or individual needs to identify all available choices and rank them by the level of welfare they provide. This process is called "evaluation."

Therefore, "evaluation" refers to a process of comparing various choices to rank them (based on a certain criterion) by order of attractiveness. At a much narrower level, the same evaluation process is performed within a health system or (even more narrowly) within a segment of a health system (e.g., hospitals), or even within the methods of management of a single disease (available budget, treatment options, maximization of patient/taxpayer welfare, etc.). In this book we deal extensively with the criteria we use to make decisions in the health care industry.

The definition of economic evaluation we first provided includes the word "systematic." It is obvious that analyses based on simplistic criteria of cost comparison between treatments (such as the price of one product compared with another), which include neither the entire financial burden nor the associated benefit for each treatment, are not systematic. When performing economic evaluations, we seek a complete, thorough, and accurate comparison of the alternatives to judge their attractiveness and then rank them. In conclusion, we might say that economic evaluation combines objective data (prices of production factors, medical technology, etc.) with preference data to rule on which of the available options maximize the welfare of the society in general or of specific patient groups.

EVIDENCE-BASED MEDICINE AND EVIDENCE-BASED HEALTH ECONOMICS

A common misconception is that economic evaluation deals exclusively with the (albeit systematic) determination of the cost or the difference in cost between two therapeutic interventions, with no regard for their effectiveness. If that were true, then economic evaluation would be demoted to a simple "accounting" assessment of burden, which would overlook the social dimension of this science.

For a correct, ambitious, and complete analysis, this would be inaccurate. Economic evaluation of health services combines cost and benefit data to help the decision-makers evaluate them. In this context, economic evaluation is useful not only to economists but also to clinical scientists and policymakers, who can so be informed about the available options (and the consequences of adopting them for the population's health and the country's budget). In the United States and elsewhere, this combination is referred to as value. Value can be thought of as a relationship between outcomes and cost, and this relationship can be graphically represented (Figure 1.1) (Williams, 2014). Value-based health care is emerging as a central dogma for health care reform. As such, intelligent application of systematic economic evaluation becomes a necessary element.

This section requires more elucidation. Generally speaking, we are all more or less familiar with the concept of evidence-based medicine (EBM). EBM is usually synonymous with the search for knowledge (regarding effectiveness) through systematic review of the literature to identify and propose optimal practices for a health system. Its purpose is to inform the people doing clinical work about these practices and to change current practices if they are suboptimal. In formalistic terms, we might say that EBM is the foundation of medical decisions on the process of systematic search, evaluation, and implementation of the findings of current research. This "movement," which is closely linked to the principles and the

Medical and/or service outcomes	Cost of care decreased	Cost of care unchanged	Cost of care increased
Improved			
Unchanged			
Worsened			

Figure 1.1 *The value framework.* Value is represented as a relationship between cost and outcomes of care. High-value interventions are depicted in green, lower-value interventions are depicted in yellow, and interventions of unacceptable value are depicted in red. High-value interventions should be rapidly implemented. Interventions of unacceptable value should be eliminated. The yellow boxes require weighing the magnitude of the changes in outcomes and costs to determine the acceptability of the intervention to the relevant stakeholder.

philosophy of evaluation, supports the notion that EBM should draw its conclusions from outcome research (Braslow, 1999; Claridge and Fabian, 2005; Sur and Dahm, 2011).

In this way, EBM attempts to link its indications with medical practice to improve health care quality and effectiveness at a *personalized* level. The implementation of its principles requires the integration of personal clinical experience with the most reliable scientific clinical data. Historically, the method of physician self-education in practicing EBM was developed at the Canadian McMaster University, and in Europe important work is being done at Oxford, at the NHS Research and Development Centre for EBM, which is its counterpart in the United Kingdom (Cohen, 1996; Hadley et al., 2007). In contrast, the United States does not have a single center for EBM. Instead, a number of stakeholders have assumed roles, including the government (e.g., Agency for Healthcare Research and Quality [AHRQ], http://www.ahrq. gov; Health Resources and Services Administration [HRSA], http://www. hrsa.gov), public–private partnerships (Patient Centered Outcomes Research Institute [PCORI], http://www.pcori.org; United States Preventive Services Task Force [USPSTF], http://www.uspreventiveservicestaskforce.org), and professional societies.

In many cases, however, the distinction between "good clinical practice" and "available economic resources" is controversial and often provides results opposite to those initially expected. For example, the adoption of practices that offer very little therapeutic benefit but unbearable burden may undercut the overall ability of the health system to treat patients in the near future and over the next generation. In other cases, society elects—through its health care system—to transfer resources to those less fortunate, regardless of whether this transfer will achieve a small increase in overall social welfare. Such social groups are usually unable to "obtain welfare" as easily as wealthier population groups. In this sense, help for such groups is not a behavior that leads to efficiency maximization; however, society might choose to uphold the concept of equity at the cost of maximizing efficiency, a fact that may not be taken into account by EBM.

In this framework, economic evaluation attempts to link EBM, whose approach is more "clinically targeted," to the wishes of society, patients, and the state to better achieve multiple goals such as viability, societal fairness, and improved efficiency in the health system. This approach is called *evidence-based health economics* (EBHE). Today, at a theoretical level, EBHE is a basic tool for the rational distribution of resources and a central point of reference for health economics methodology. The arguments presented here might indicate that, at a purely theoretical level, EBHE is broader than EBM and contains EBM, but in practice they are equal allies with different primary goals. EBM aims to influence basic clinical practices, whereas the evaluation of health technology is aimed at supporting the exercise of health policies (Banta, 2003).

THE NECESSITY OF DRAFTING ECONOMIC EVALUATIONS

Generally speaking, one may say that the goal of a health system is to provide high-quality health services to their defined population on an equal basis, to allow quick access to innovation that improves value, to produce a large number of health services to meet the needs of the population, and to do this efficiently by consuming as few resources as possible. These health system activities are centrally organized through complex systems of political oversight, planning, and financing.

Nevertheless, achieving these goals is impeded by certain factors, the most important of which are the following:
- The demographic problem (few active workers to support the system but many retirees in societies where health care funding is tied to employment)

- The resolution to provide full health care coverage to the population (especially in Western societies), which increases costs
- The modern unhealthy lifestyle (carbohydrate-rich foods, sedentary lifestyle, use of alcohol, smoking, lack of exercise, poor diet, excessive consumption of drugs, etc.), which causes chronic diseases and complications that can only be treated at significant costs and with modest therapeutic results
- The financial costs imposed by the technological advances in health services (businesses bear very high research and development costs that they wish to transfer to the end consumer or the public insurance funds while also making some profit because they are profit-seeking enterprises), many of which add little incremental benefit and result in poor value
- The extended average lifespan (older people suffer from multiple conditions and chronic diseases, with high treatment costs)
- The public's expectations (it is the express conviction of a democratic state that the citizens' needs must be met with no particular consideration of the cost; furthermore, the citizens' demands have increased because of improved educational systems and cheap and widely available communication/information channels such as the Internet, among others)
- Medical errors (which can be harmful or fatal for the patient): it is estimated that medical errors cost billions of euros or other currency and thousands of lives each year (van den Bos et al., 2011)
- Unnecessary consumption of health care resources (supply induced)

Because all of these factors constitute a direct or indirect financial burden on modern health systems, governments believe that the money spent for health care is excessive and that priorities must be set, or the ratio at which the state and the patient share this expense must be altered. Economic evaluation allows a more rational process to evaluate the factors that impact the system. We should underline that an absolute restriction of health care expenditures is a difficult goal from a social point of view, whereas a reduction in the rate of expenditure growth is easier to achieve (so-called bending of the cost curve). Therefore, the aim of economic evaluation is not necessarily to restrict health care expenditures, but rather to rationally distribute the available resources to those uses that will achieve the highest possible level of population health based on certain societal criteria. In certain cases, such criteria may lead to an increase in expenditures when this is financially or socially acceptable (yellow box in web version of Figure 1.1). If the ultimate goal were to reduce expenditures, then the state would simply cease to provide health care services to certain citizens, which would achieve immediate savings but would inflame the public sense of justice and would strain social cohesion. Such expenditure restrictions are socially justified and financially effective only if they include a substantial restructuring of the system to save resources without reducing benefits.

In conclusion, economic evaluation is more of a guideline for a systematic and strict way of thinking than a dogmatic set of quantification methods that will mechanically determine the therapeutic options to which society should direct its resources.

REFERENCES

Banta, D., 2003. The development of health technology assessment. Health Policy 63 (2), 121−132.

Braslow, J.T., 1999. History and evidence-based medicine: lessons from the history of somatic treatments from the 1900s to the 1950s. Ment. Health Serv. Res. 1 (4), 231−240.

Claridge, J.A., Fabian, T.C., 2005. History and development of evidence-based medicine. World J. Surg. 29 (5), 547−553.

Cohen, L., 1996. McMaster's pioneer in evidence-based medicine now spreading his message in England. Can. Med. Assoc. J. 154 (3), 388−390.

Hadley, J.A., Davis, J., Khan, K.S., 2007. Teaching and learning evidence-based medicine in complementary, allied, and alternative health care: an integrated tailor-made course. J. Altern Complement Med. 13 (10), 1151−1155.

Sur, R.L., Dahm, P., 2011. History of evidence-based medicine. Ind. J. Urol. 27 (4), 487−489.

Van Den Bos, J., Rustagi, K., Gray, T., Halford, M., Ziemkiewicz, E., Shreve, J., 2011. The $17.1 billion problem: the annual cost of measurable medical errors. Health Aff. (Millwood) 30 (4), 596−603.

Williams, M.S., 2014. Genomic medicine implementation: learning by example. Am. J. Med. Genet. C Semin. Med. Genet. 166, 8−14.

CHAPTER 2

Genomic Medicine Today: An Introduction for Health Economists

INTRODUCTION

The central dogma of genomic medicine is to exploit the individual's genomic sequence in combination with other clinical information to enhance the clinical decision-making process. In recent years, significant advances have enriched our knowledge regarding the molecular etiology of a wide range of human genetic diseases, allowing for better disease prevention, prognosis, and treatment. In parallel, genomic technology has progressed rapidly and, as a result, genomics research has the potential to aid clinicians in their task of estimating disease risk and individualizing treatment modalities. This constitutes the basis of genomic medicine, a new discipline that promises to enhance opportunities for the customization of patient care and the personalization of conventional therapeutic interventions.

The concept of genomic or personalized medicine stems from as early as 400 B.C., at which time Hippocrates of Kos (460–370 B.C.E.) stated that "...it is more important to know what kind of person suffers from a disease than to know the disease a person suffers." This ancient statement encapsulates the essence of modern personalized genomic medicine. The first application of individualized care based on presumed genetic information is documented in the Talmud (Yevamot 64b). Rabbi Judah the Prince (135–217 C.E.) ruled that if a woman's first two children died from blood loss after circumcision, the third son should not be circumcised. A second disagreed and ruled that the third son may be circumcised; however, if this infant died, the fourth child should not be circumcised. There was agreement that abnormal bleeding was hereditary but disagreement regarding how many events were required to establish a pattern and therefore exempt a child from circumcision (http://download.yutorah.org/2013/1053/793670.pdf). At present, there is a significant effort worldwide to support the translation of genomic research into clinical practice so that genomic medicine can ultimately be used to benefit the global community.

PHARMACOGENOMICS AND GENOMIC MEDICINE

For more than 50 years, it has been known that there is substantial inter-individual variability in the clinical response to drug treatments and that only a percentage of patients respond satisfactorily to their medications. This implies that the balance of the patient population either is not receiving proper medication or, worse, is suffering from either adverse drug reactions, varying from mild to even lethal, or marked therapeutic delays, which are caused by empiric trials of different medications until an acceptable alternative is identified (Spear et al., 2001). Furthermore, the side effects for the same drug vary from patient to patient. Experimental evidence suggests that the variable phenotypic expression of drug treatment efficacy and toxicity is determined by a complex interplay of multiple genetic variants and environmental factors (Squassina et al., 2010).

Pharmacogenomics is referred to as "...the delivery of the right drug to the right patient at the right dose" (Piquette-Miller and Grant, 2007), and several pharmacogenomic testing approaches currently exist to identify the underlying pharmacogenomic biomarkers. The availability of knowledge about the impact of pharmacogenomics variants on drug response is the reason that pharmacogenomics is the branch of genomic medicine that is being actively implemented in many centers.

APPLICATIONS OF PHARMACOGENOMICS

At present, the US Food and Drug Administration (FDA) and the European Medicines Agency (EMA) provide pharmacogenomic information on the label of more than 150 drugs. Although only a few recommend that pharmacogenomic tests must be performed prior to prescribing these drugs, the inclusion of the information for providers is indicative of the importance these agencies place on this emerging field. The Clinical Pharmacogenetics Implementation Consortium has published a series of evidence-based clinical guidelines for use of pharmacogenetic information to guide therapy (http://www.pharmgkb.org/page/cpic). Currently, pharmacogenomics instances are most relevant to the specialties of oncology, cardiology, and psychiatry, and they are emerging for several other medical specialties.

Oncology Examples

Trastuzumab, a monoclonal antibody blocking v–erb–b2 erythroblastic leukemia viral oncogene homolog 2 (HER2, also ERBB2) receptor protein, is one of the prime examples for which pharmacogenomic testing is routinely used. It has been shown that variable expression of the *HER2* receptor gene determines the likelihood that a patient will respond to trastuzumab. *HER2* overexpression is observed in approximately one-fourth of breast cancer patients. It is correlated with poor prognosis, increased tumor formation, and metastasis, as well as resistance to chemotherapeutic agents. As such, *HER2* testing either through genetic testing or through analysis of the HER2 protein product identifies patients who will likely respond to trastuzumab. Other examples are the tyrosine kinase inhibitors erlotinib and gefitinib, which are designed to target the epidermal growth factor receptor (EGFR). EGFR has been shown to play a role in predisposition to lung cancer. *EGFR* variants are now used to predict improved progression-free survival with gefitinib in a comparison with carboplatin—paclitaxel (Mok et al., 2009).

Irinotecan is yet another example of a drug for which a pharmacogenomic test can help to identify colorectal cancer patients who are likely to experience adverse reactions during treatment, such as diarrhea and severe neutropenia. The *UGT1A1*28* variant is associated with increased toxicity to irinotecan; as such, patients homozygous for the *UGT1A1*28* allele are at higher risk for development of irinotecan-associated neutropenia and diarrhoea. The FDA recommended an addition to the irinotecan package insert to include *UGT1A1*28* genotype as a risk factor for severe neutropenia. However, it is important to note that in the Evaluation of Genomic Applications in Practice and Prevention (EGAPP) working group's recommendation regarding irinotecan, it was noted that there is evidence that presence of this genotype results in improved response to irinotecan, that is, improved efficacy (Berg et al., 2009). This means that when evaluating a pharmacogenomic test, care must be taken to identify impact on both efficacy and adverse events. The following example also illustrates this principle.

Azathioprine, 6-mercaptopurine, and 6-thioguanine are drugs that are metabolized by the thiopurine methyltransferase (TPMT) enzyme. They are primarily used for the treatment of acute lymphoblastic leukemia (ALL), but they are also used for inflammatory conditions such as Crohn's disease, ulcerative colitis, and psoriasis. Patients with inherited TPMT deficiency have been shown to be likely to experience severe (potentially fatal) hematopoietic toxicity when exposed to standard doses of thiopurine drugs. As such, a pharmacogenomic test could classify patients according to normal, intermediate, and deficient levels of TPMT activity, which enables physicians to prescibe the correct dose to these patients based on their TPMT activity levels. In other words, patients classified as having normal activity are treated with conventional doses, whereas lower doses (from 50% [intermediate metabolizers] to as low as 10% [poor metabolizers] of the normal dose) are administrated to avoid toxicity in deficient and intermediate patients who are likely to suffer exaggerated, potentially life-threatening toxic responses to normal doses of azathioprine and thiopurine drugs (Relling et al., 1999). Again, there is evidence that individuals with intermediate or poor metabolizers may have improved clinical response (efficacy) to these agents (Stanulla et al., 2005). This is less of an issue in ALL because the endpoint of treatment is measured by carefully monitoring end organ response (white count, minimal residual disease, etc.). This demonstrates another important principle: if you can measure a relevant phenotype, that is preferable to using a surrogate endpoint such as a genotype. So in the case of ALL, although the starting dose of the medication could be chosen based on genotype, the doses will be adjusted based on the phenotypic response of the bone marrow. This is likely quite different for other treatment indications such as inflammatory bowel disease (IBD) or psoriasis. In these conditions, white blood count is not measured as an indicator of treatment response; therefore, there is higher risk of missing an incipient bone marrow catastrophe in individuals with intermediate or poor metabolizer phenotypes. As a consequence, the importance of the pharmacogenomic information could be more important in these conditions. This has led to development of a therapeutic algorithm for use of thiopurines in IBD (Chouchana et al., 2012).

Cardiology Examples

In recent years, cardiology became the other medical specialty in which pharmacogenomics applications are emerging into practice, particularly related to warfarin, acenocoumarol, and clopidogrel. For more than half a century, coumarinic oral anticoagulants (COAs), namely warfarin, acenocoumarol, and phenprocoumon, have been the standard oral anticoagulants for thromboembolic disorders. These COAs, however, have a narrow therapeutic window and are associated with high risk of major bleeding, especially during the initial phase of treatment, making them one of the leading causes of emergency hospitalizations worldwide. Experimental evidence suggests that there is a substantial interindividual variation in COA treatment response, which requires regular monitoring and dosage adjustment (Pirmohamed, 2006).

It has been shown for many years that *CYP2C9* and *VKORC1* genomic variations influence interindividual the variable response of COAs. CYP2C9 metabolizes COA, and vitamin K epoxide reductase (VKORC1) is the pharmacologic target enzyme of these drugs (Bodin et al., 2005; Schalekamp et al., 2007). The *CYP2C9*2* and **3* variants lead to decreased CYP2C9 activity, whereas the *VKORC1 − 1639G > A* variant influences pharmacodynamic response to coumarins. Based on these findings, there are several dosing algorithms incorporating *CYP2C9* and *VKORC1* genotype information to predict the patient's response to warfarin (Gage, et al., 2008; Klein et al., 2009). Warfarin is among the drugs that have been relabeled, as early as in 2007, to include *CYP2C9* and *VKORC1* genotyping as a means to optimize warfarin dosing.

Clopidogrel is another drug that is the standard for the care of acute coronary syndromes for which nonresponsiveness, corresponding to approximately 25% of patients receiving clopidogrel, is related to recurrent ischemic events (Gladding et al., 2008). Response to clopidogrel has been shown to be determined by the *CYP2C19* genotype (Geisler et al., 2008) because clopidogrel is a pro-drug that requires enzymatic conversion to an active ingredient. CYP2C19 is the primary cytochrome P450 enzyme for this conversion. The *CYP2C19*2* allele has been shown to impair CYP2C19 function, and this is associated with a marked decrease in patient responsiveness to clopidogrel (Mega et al., 2009).

Psychiatric Examples

Unlike oncology and cardiology, pharmacogenomics and personalized medicine are still far from being achieved in psychiatry (Alda, 2013). Pharmacokinetic pathways in which cytochrome P450 isoenzymes, such as CYP2D6, CYP1A1, CYP3A4, CYP2C9, and CYP2C19, are involved may predict serum levels of antidepressants and antipsychotics. Numerous studies implicated *CYP2D6* genomic variants as a predictor of risperidone-induced adverse reactions, but not those of clozapine (reviewed in Tsermpini et al., 2014). A recommendation by the EGAPP working group found insufficient evidence to support a recommendation for or against use of CYP450 genetic testing in adults beginning selective serotonin reuptake inhibitor (SSRI) treatment for nonpsychotic depression (EGAPP, 2007). As far as atypical antipsychotics are concerned, pharmacogenomic studies have mostly focused on the *HTR2A* and *HTR2C* serotonin receptor genes (Arranz and de Leon, 2007) and their role in SSRIs. Finally, there are few genetic determinants that have been recently reported to predict response to lithium chloride treatment of bipolar disorder (Perlis et al., 2009a; Squassina et al., 2011).

Infectious Disease Examples

Last but not least, pharmacogenomics for infectious diseases is an expanding area that is gradually assuming an important role. One example is prediction of adverse effects caused by antiretroviral drug therapies that are used in the treatment of HIV/AIDS and chronic hepatitis C (Picard and Bergeron, 2002). Mallal et al. (2008) demonstrated that the HLA-B*5701 allele is indicative of a hypersensitive reaction to abacavir. The latter variant displays varying frequencies in patients from different populations and reaches high frequencies in Asian populations, making HLA-B*5701 an important pharmacogenomics marker to predict abacavir-induced toxicity. Genetic variation can also inform differences in response to treatment. An example of this is variation in the interleukin gene *IL28B* in patients being treated for chronic hepatitis C (Bota et al., 2013). This is used in an economic analysis in Chapter 6.

The astute reader will note that the examples provided lend themselves to the tools of economic analysis that will be introduced in the next two chapters. Several of these examples are the subject of application of economic analysis and appear later in the book.

FACING THE CURRENT CHALLENGES OF GENOMIC MEDICINE

To maximize the contribution of genomic medicine in current medical practice, there are certain actions that should be taken and obstacles to be overcome to expedite the integration of pharmacogenomics into modern medical practice and, hence, to maximally exploit the benefits of this new discipline for society (Figure 2.1). In the following

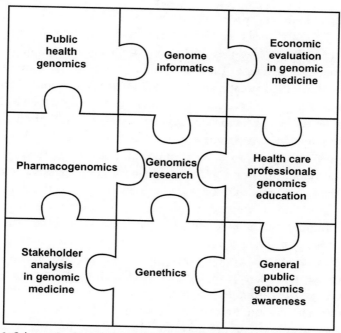

Figure 2.1 Schematic depiction, as a jigsaw puzzle, of the various disciplines that constitute the multidisciplinary field of genomic medicine. Note the central role of genomics research, which is depicted as the central piece in the puzzle that needs to be complemented by the other pieces to fulfill the promise of genomic medicine.

paragraphs, we outline the various challenges and bottlenecks that should be overcome and steps that have to be undertaken for pharma-cogenomics to find its place in future medicine.

EDUCATION OF HEALTH PROFESSIONALS

A prior requirement for the introduction of pharmacogenomics in some medical specialties is to ensure high-level genomics education for health care professionals. An important bottleneck is to fill the gap between the pharmacogenomic testing *per se* and the interpretation and utilization of its results in a clinical setting, also known as translation of genomic results into patient care (Patrinos, 2010; Reydon et al., 2012). In this context, several international organizations have called for the integration of pharmacogenomics and personalized medicine education in core medical curricula (Gurwitz et al., 2003). Several medical, pharmacy, and health science curricula are gradually integrating such courses for their under-graduate and postgraduate students (Pisanu et al., 2014). Moreover, the urgent need to incorporate genetic and pharmacogenomics education into medical and pharmacy schools has been motivated by the challenging ethical implications of personalized medicine (Frueh and Gurwitz, 2004).

At the same time, researchers in the field of public health genomics are gradually starting to assess the attitudes of health care professionals and the general public regarding pharmacogenomics and genomic medicine, because these opinions may be used to harmonize the development of genomic medicine, molecular diagnostics, and related disciplines with those of other countries, establish the legal framework to control genetic testing services, safeguard the interests of the citizens, and shape the over-all genomic medicine policy environment. There are several studies indicating that patients and the general public expect to receive pharma-cogenomic services from health care professionals who can confidently explain the test and interpret its implications for prescriptions. However, respondents of these surveys indicate that a gap still exists between patients' high expectations and health care professionals' knowledge (Mai et al., 2011, 2014), although both stakeholders hold a very positive view of genomic medicine (Mitropoulou et al., 2015). Interestingly, a stake-holders' awareness survey that queried stakeholders (academia, health care professionals, industry, government) involved in pharmacogenomics in Japan evidenced the same pattern of expectations, although they have also expressed similar concerns regarding issues such as poor genetic awareness from the general public and the possibility of breaching privacy (Tamaoki et al., 2007). Although the level of pharmacogenomic education among health care professionals is still not optimal, proper interventions at the educational level will definitely have a very positive effect on increasing genetics education, which would consequently facilitate the incorporation of genetics into patient care.

DEMONSTRATION OF CLINICAL UTILITY

A key to implementation of a new technology in practice is evidence that the new technology makes a difference. Clinical utility means assessing the balance of the benefits and risks of a given technology. In its narrow-est sense, clinical utility refers to the ability of a screening or diagnostic test to prevent or ameliorate adverse health outcomes such as mortality, morbidity, or disability through the adoption of efficacious treatments on the condition of test results. Simply, if the benefits outweigh the risks based on robust evidence, a technology is said to have clinical utility. In 1997, the National Institutes of Health—Department of Energy Task Force on Genetic Testing stated, "Before a genetic test can be generally accepted in clinical practice, data must be collected to demonstrate the benefits and risks that accrue from both positive and negative results" (Holtzman and Watson, 1997). Evidence of clinical utility, with a few exceptions, has been lacking, as evidenced by the majority of EGAPP working group recommendation statements concluding that there is insufficient evidence to recommend for or against use of a given genetic test for a certain defined purpose. Evidence of clinical utility is ultimately translated into guidelines that can effectively alter clinical practice when effectively implemented. The lack of guidelines in genomic medicine is a key contributor to its general lack of integration into practice. The afore-mentioned CPIC guidelines are an intentional effort to address this lack of evidence in the area of pharmacogenomics.

HEALTH CARE COSTS

Generally speaking, the cost-effectiveness of a pharmacogenomic test is a crucial factor for its adoption in a clinical setting. Thus, to demonstrate its economic benefits, pharmacogenomic testing requires evidence not only of its clinical utility but also of its cost-effectiveness (Deverka et al., 2010). Regarding their impact on different medical specialties, there are not many reports indicating that certain drug treatments related to psychiatry (Perlis et al., 2009b), cancer (Carlson et al., 2009), and chronic inflammatory diseases (Priest et al., 2006) are cost-effective. For example, Perlis et al. (2009b) have demonstrated cost-effectiveness and benefits of a putative pharmacogenomic test for antidepressant efficacy in the treatment of a major depressive disorder. It should be noted, however, that for this as well as for every pharmacogenomic test, an equally important factor to determine cost-effectiveness is the pharmacogenomic marker allele frequencies that vary significantly among different populations. The most characteristic examples of the application of economic evaluation in pharmacogenomics and genomic medicine are described in detail in Chapter 6.

Different reimbursement policies of the various health care systems, particularly in developing countries with large fiscal deficits are another important parameter for health care costs related to genomic medicine (Mette et al., 2012). In other words, payers, particularly public insurance funds, can obviate the rapid dissemination of pharmacogenomics (Ginsburg and Willard, 2009; Mitropoulou et al., 2015). Conversely, in the United States, the fragmented nature of the delivery and reimbursement systems leads to different groups making different decisions, even when faced with the same evidence. Williams (2007) provides a comprehensive review of this problem in pharmacogenomics. Overall, pharmacogenomic testing and genomic medicine in general may represent a valuable resource for health care decision-makers, leading to increased quality of clinical care and increased economic benefits for the pharmaceutical industry and public health, although pharmacogenomic tests will probably prove to be more cost-effective than cost-saving, particularly for certain disease/treatment modalities and populations (Deverka and McLeod, 2008; Flowers and Veenstra, 2004; Mette et al., 2012).

INSURANCE AND PRIVACY ISSUES

Considering the fact that pharmacogenomics and genomic medicine strongly rely on interindividual genomic variability, these disciplines have to deal with a number of issues related to genetic discrimination/stigmatization, privacy, and possible implications for access to life and health insurance. In 1998, Matloff et al. (2000) surveyed the members of the National Society of Genetic Counselors (NSGC) Special Interest Group (SIG) regarding cancer and demonstrated that more than half of the respondents would not bill their insurance companies for genetic testing, largely because of fear of genetic discrimination. Similarly, in 2007, Huizenga et al. (2010), compared the results of a new survey provided to NSGC SIG using the data of Matloff et al. (2000) and pointed out a notable change in perceptions and behavioral intent among cancer genetics professionals (CGPs) over time, indicating that fear of genetic discrimination has become significantly less common since 1998.

Although the data presented concern the broad concept of genetic testing, pharmacogenomics has to deal with the same issues of genetic stigmatization/discrimination (Robertson, 2001; Issa, 2002), particularly because the social consequences arising from new disease labels, namely "good, poor, or nonresponder" to a given treatment modality, would touch on very sensitive issues of interpersonal stigmatization or identity. Moreover, pharmaceutical companies could voluntarily ignore, for economic reasons, patients with rare or complex genetic conditions or those who are not responding to any known treatment, leading to consequent deprivation of effective treatments for certain subpopulations (Rothstein and Epps, 2001).

Also, an equally important issue that may raise concern, particularly in the whole-genome sequencing era, is the storage of genomic information in databases with the inherent danger of losing confidentiality or privacy, because an enormous quantity of genotypic, phenotypic, and demographic data regarding individuals are inter-related (Vaszar et al., 2003; Potamias et al., 2014). To this end, protection of privacy and confidentiality has to be ensured, particularly in the whole-genome sequencing era, because pharmacogenomic tests can carry several types of secondary information that represent a risk of psychosocial harm and adverse insurance and/or employment implications. Moreover, particular subpopulation groups may face genetic discrimination when trying to access health care or health insurance (Smart et al., 2004). The Genetic Information Non-discrimination Act (GINA), the Health Insurance Portability and Accountability Act Privacy Rule, and the Genomics and Personalized Medicine Act (GPMA) are all legislative attempts in the United States trying to address these questions, although they often lack clarity in critical issues regarding translation of human genetic variation from the bench to the bedside (Lee and Mudaliar, 2009).

CONCLUSIONS AND FUTURE PERSPECTIVES

The dawn of the postgenomics era has further revealed the role of pharmacogenomics and genomic medicine as two of the most fundamental disciplines of modern clinical practice toward achieving the goal of personalized medicine. As with every new trend, these new disciplines have found their way faster into some medical specialties, such as cardiology or oncology, in which they are already applied for selecting and/or dosing a specific medication, whereas in other specialties, such as psychiatry or infectious diseases, they are still lagging behind. Even though personalized medicine is not yet a widespread practice, ongoing international efforts confirm that this concept is close to becoming a reality (Kampourakis et al., 2014).

The delineation of genomic variation with complex pathways pertaining to drug efficacy and toxicity and the etiology of complex genetic traits and clinical conditions constitute a daunting challenge, because these phenotypes are, by their nature, more complex and heterogeneous when compared with bimodal traits. Although this may sound discouraging, there are some strategies that, if used correctly, may lead us to translate genomics research findings into genomic medicine applications.

In the whole-genome sequencing era, DNA sequence analysis allows the identification of unique or very rare novel single nucleotide polymorphisms (SNPs) either in genes encoding for drug-metabolizing enzymes and transporters or in the whole genome with direct implications in drug response (Mizzi et al., 2014). Preliminary evidence suggests that there are several thousand variants in each individual affecting drug response, hundreds of which are novel and whose identification is of utmost importance in determining one's personalized pharmacogenomics profile (Mizzi et al., 2014). Moreover, some analytical approaches, such as case-only genome-wide interaction (COGWI) (Pierce and Ahsan, 2010), can provide a straightforward method for detecting pharmacogenomic interactions related to treatment response in complex diseases, which might provide a framework of operational and decisional criteria for setting the course leading to widespread use of pharmacogenomics in public health.

Furthermore, there are novel concepts that emerge in the field of pharmacogenomics, such as some mitochondrial variants that could be considered putative pharmacogenomic markers (Pacheu-Grau et al., 2010), or some genomic variants that could be correlated with variable responses to drug therapy in organ transplantation. In the latter case, tacrolimus constitutes perhaps the best established example of the *CYP3A5* gene effect on drug disposition and dosage, with the *CYP3A5*3* allele shown to be associated with a lower dose of tacrolimus to achieve therapeutic blood concentrations. There are also other drugs used in organ transplantation whose effectiveness has been correlated with a number of genomic variants, such as azathioprine with *TPMT* alleles, cyclosporine with *ABCB1* alleles, sirolimus (like tacrolimus) with *CYP3A5* alleles, and corticosteroids with *ABCB1* and *IL10* polymorphisms (reviewed by Girnita et al., 2008).

Several international organizations, policy and advocacy groups, and research consortia have been formed, such as the Personalized Medicine Coalition (PMC; http://www.personalizedmedicinecoalition.org; Abrahams et al., 2005) and the European Alliance for Personalized Medicine (EAPM; http://www.eapm.eu), which are nonprofit policy and advocacy groups formed of pharmaceutical biotechnology, diagnostic and information technology companies, health care providers and payers, patient advocacy groups, industry policy organizations, academic institutions, and government agencies, aiming to facilitate the use of personalized medicine approaches, to help in the training of the public; to provide opinion leadership; and to convey information to the media, government officials, and health care leaders. Other initiatives, such as the Genomic Medicine Alliance (http://www.genomicmedicinealliance.org), a newly established international research consortium (Cooper et al., 2014), aim specifically to bridge the gap between developed and developing nations by ensuring technology and knowledge transfer. Such efforts would be particularly useful for developing nations to defray health care costs and improve quality of life by minimizing adverse drug reactions (Mitropoulos et al., 2011).

From the time of Hippocrates, personalized medicine has been considered a component of good medical practice. However, as with every new trend, extra caution should be taken to avoid abuse of or underestimating the potential of this new technology. It should be remembered that genomic information is only one type of information that may be relevant to a clinical situation and that it does not occupy a privileged position in relation to other types of information. The importance of understanding the patient's goals for treatment has not been given as much weight as it deserves. This "patient-centered" approach is being increasingly emphasized but it is not new. Sir William Osler, another icon of medicine, recognized its importance in care when he said, "Care more for the individual patient than for the special features of the disease...Put yourself in his place. ..." We therefore favor the definition of personalized medicine proposed by Pauker and Kassirer in 1987 that embraces the concept of patient centeredness and equality of information: "Personalized medicine is the practice of clinical decision-making such that the decisions made **maximize the outcomes that the patient most cares about and minimizes those that the patient fears the most**, on the basis of **as much knowledge about the individual's state** as is available."

In conclusion, the adoption of pharmacogenomics and genomic medicine for the prescription of personalized medical interventions still has to face a number of challenges, but the increasing knowledge of the molecular basis of inherited conditions and of drug efficacy and toxicity, accompanied by the growing attention of the pharmaceutical industry and national health care policymakers, will probably accelerate the pace toward the achievement of personalized health care and precision medicine.

REFERENCES

Abrahams, E., Ginsburg, G.S., Silver, M., 2005. The personalized medicine coalition: goals and strategies. Am. J. Pharmacogenomics 5, 345–355.

Alda, M., 2013. Personalized psychiatry: many questions, fewer answers. J. Psychiatry Neurosci. 38, 363–365.

Arranz, M.J., de Leon, J., 2007. Pharmacogenetics and pharmacogenomics of schizophrenia: a review of last decade of research. Mol. Psychiatry 12, 707–747.

Berg, A.O., Armstrong, K., Botkin, J., Calonge, N., Haddow, J., Hayes, M., et al., 2009. Recommendations from the EGAPP working group: can UGT1A1 genotyping reduce morbidity and mortality in patients with metastatic colorectal cancer treated with irinotecan? Genet. Med. 11 (1), 15–20.

Bodin, L., Verstuyft, C., Tregouet, D.A., Robert, A., Dubert, L., Funck-Brentano, C., et al., 2005. Cytochrome P450 2C9 (CYP2C9) and vitamin K epoxide reductase (VKORC1) genotypes as determinants of acenocoumarol sensitivity. Blood 106, 135–140.

Bota, S., Sporea, I., Şirli, R., Neghină, A.M., Popescu, A., Străin, M., 2013. Role of interleukin-28B polymorphism as a predictor of sustained virological response in patients with chronic hepatitis C treated with triple therapy: a systematic review and meta-analysis. Clin. Drug Investig. 33 (5), 325–331.

Carlson, J.J., Garrison, L.P., Ramsey, S.D., Veenstra, D.L., 2009. The potential clinical and economic outcomes of pharmacogenomic approaches to EGFR-tyrosine kinase inhibitor therapy in non-small-cell lung cancer. Value Health 12, 20–27.

Chouchana, L., Narjoz, C., Beaune, P., Loriot, M.A., Roblin, X., 2012. Review article: the benefits of pharmacogenetics for improving thiopurine therapy in inflammatory bowel disease. Aliment. Pharmacol. Ther. 35 (1), 15–36.

Cooper, D.N., Brand, A., Dolzan, V., Fortina, P., Innocenti, F., Lee, M.T., et al., 2014. Bridging genomics research between developed and developing countries: the genomic medicine alliance. Pers Med. 11, 615–623.

Deverka, P.A., McLeod, H.L., 2008. Harnessing economic drivers for successful clinical implementation of pharmacogenetic testing. Clin. Pharmacol. Ther. 84, 191–193.

Deverka, P.A., Vernon, J., McLeod, H.L., 2010. Economic opportunities and challenges for pharmacogenomics. Annu. Rev. Pharmacol. Toxicol. 50, 423–437.

Evaluation of Genomic Applications in Practice and Prevention (EGAPP) Working Group, 2007. Recommendations from the EGAPP working group: testing for cytochrome P450 polymorphisms in adults with nonpsychotic depression treated with selective serotonin reuptake inhibitors. Genet. Med. 9 (12), 819–825.

Flowers, C.R., Veenstra, D., 2004. The role of cost-effectiveness analysis in the era of pharmacogenomics. Pharmacoeconomics 22, 481–493.

Frueh, F.W., Gurwitz, D., 2004. From pharmacogenetics to personalized medicine: a vital need for educating health professionals and the community. Pharmacogenomics 5, 571–579.

Gage, B.F., Eby, C., Johnson, J.A., Deych, E., Rieder, M.J., Ridker, P.M., et al., 2008. Use of pharmacogenetic and clinical factors to predict the therapeutic dose of warfarin. Clin. Pharmacol. Ther. 84, 326–331.

Geisler, T., Schaeffele, E., Dippon, J., Winter, S., Buse, V., Bischofs, C., et al., 2008. CYP2C19 and nongenetic factors predict poor responsiveness to clopidogrel loading dose after coronary stent implantation. Pharmacogenomics 9, 1251–1259.

Ginsburg, G.S., Willard, H.F., 2009. Genomic and personalized medicine: foundations and applications. Transl. Res. 154, 277–287.

Girnita, D.M., Burckart, G., Zeevi, A., 2008. Effect of cytokine and pharmacogenomic genetic polymorphisms in transplantation. Curr. Opin. Immunol. 20, 614–625.

Gladding, P., Webster, M., Ormiston, J., Olsen, S., White, H., 2008. Antiplatelet drug nonresponsiveness. Am. Heart J. 155, 591–599.

Gurwitz, D., Weizman, A., Rehavi, M., 2003. Education: teaching pharmacogenomics to prepare future physicians and researchers for personalized medicine. Trends Pharmacol. Sci. 24, 122–125.

Holtzman, N.A., Watson, M.S. (Eds.), 1997. Promoting safe and effective genetic testing in the United States. Final report of the task force on genetic testing. <http://www.genome.gov/10001733>.

Huizenga, C.R., Lowstuter, K., Banks, K.C., Lagos, V.I., Vandergon, V.O., Weitzel, J.N., 2010. Evolving perspectives on genetic discrimination in health insurance among health care providers. Fam. Cancer 9 (2), 253–260.

Issa, A.M., 2002. Ethical perspectives on pharmacogenomic profiling in the drug development process. Nat. Rev. Drug Discov. 1, 300–308.

Kampourakis, K., Vayena, E., Mitropoulou, C., Borg, J., van Schaik, R.H., Cooper, D.N., et al., 2014. Key challenges for next generation pharmacogenomics. EMBO Rep. 15 (5), 472–476.

Klein, T.E., Altman, R.B., Eriksson, N., Gage, B.F., Kimmel, S.E., Lee, M.T., et al., 2009. Estimation of the warfarin dose with clinical and pharmacogenetic data. N. Engl. J. Med. 360, 753–764.

Lee, S.S., Mudaliar, A., 2009. Medicine. Racing forward: the genomics and personalized medicine act. Science 323, 342.

Mai, Y., Koromila, T., Sagia, A., Cooper, D.N., Vlachopoulos, G., Lagoumintzis, G., et al., 2011. A critical view of the general public's awareness and physicians' opinion of the trends and potential pitfalls of genetic testing in Greece. Pers. Med. 8, 551–561.

Mai, Y., Mitropoulou, C., Papadopoulou, X.E., Vozikis, A., Cooper, D.N., van Schaik, R.H., et al., 2014. Critical appraisal of the views of healthcare professionals with respect to pharmacogenomics and personalized medicine in Greece. Pers. Med. 11 (1), 15–26.

Mallal, S., et al., 2008. HLA-B*5701 screening for hypersensitivity to abacavir. N. Engl. J. Med. 358, 568–579.

Matloff, E.T., Shappell, H., Brierley, K., Bernhardt, B.A., McKinnon, W., Peshkin, B.N., 2000. What would you do? Specialists' perspectives on cancer genetic testing, prophylactic surgery, and insurance discrimination. J. Clin. Oncol. 18, 2484–2492.

Mega, J.L., Close, S.L., Wiviott, S.D., Shen, L., Hockett, R.D., Brandt, J.T., et al., 2009. Cytochrome p-450 polymorphisms and response to clopidogrel. N. Engl. J. Med. 360, 354–362.

Mette, L., Mitropoulos, K., Vozikis, A., Patrinos, G.P., 2012. Pharmacogenomics and public health: implementing populationalized medicine. Pharmacogenomics 13 (7), 803–813.

Mitropoulos, K., Johnson, L., Vozikis, A., Patrinos, G.P., 2011. Relevance of pharmacogenomics for developing countries in Europe. Drug Metabol. Drug Interact. 26, 143–146.

Mitropoulou, C., Mai, Y., van Schaik, R.H., Vozikis, A., Patrinos, G.P., 2015. Documentation and analysis of the policy environment and key stakeholders in pharmacogenomics and genomic medicine in Greece. Public Health Genomics 17, 280–286.

Mizzi, C., Mitropoulou, C., Mitropoulos, K., Peters, B., Agarwal, M.R., van Schaik, R.H., et al., 2014. Personalized pharmacogenomics profiling using whole genome sequencing. Pharmacogenomics 15 (9), 1223–1234.

Mok, T.S., Wu, Y.L., Thongprasert, S., Yang, C.H., Chu, D.T., Saijo, N., et al., 2009. Gefitinib or carboplatin–paclitaxel in pulmonary adenocarcinoma. N. Engl. J. Med. 361, 947–957.

Pacheu-Grau, D., Gomez-Duran, A., Lopez-Perez, M.J., Montoya, J., Ruiz-Pesini, E., 2010. Mitochondrial pharmacogenomics: barcode for antibiotic therapy. Drug Discov. Today 15, 33–39.

Patrinos, G.P., 2010. General considerations for integrating pharmacogenomics into mainstream medical practice. Hum. Genomics 4, 371–374.

Pauker, S.G., Kassirer, J.P., 1987. Decision analysis. N. Engl. J. Med. 316 (5), 250–258.

Perlis, R.H., Smoller, J.W., Ferreira, M.A., McQuillin, A., Bass, N., Lawrence, J., et al., 2009a. A genome-wide association study of response to lithium for prevention of recurrence in bipolar disorder. Am. J. Psychiatry 166, 718–725.

Perlis, R.H., Patrick, A., Smoller, J.W., Wang, P.S., 2009b. When is pharmacogenetic testing for antidepressant response ready for the clinic? A cost-effectiveness analysis based on data from the STAR*D study. Neuropsychopharmacology 34, 2227–2236.

Picard, F.J., Bergeron, M.G., 2002. Rapid molecular theranostics in infectious diseases. Drug Discov. Today 7, 1092–1101.

Pierce, B.L., Ahsan, H., 2010. Case-only genome-wide interaction study of disease risk, prognosis and treatment. Genet. Epidemiol. 34, 7–15.

Piquette-Miller, M., Grant, D.M., 2007. The art and science of personalized medicine. Clin. Pharmacol. Ther. 81, 311–315.

Pirmohamed, M., 2006. Warfarin: almost 60 years old and still causing problems. Br. J. Clin. Pharmacol. 62, 509–511.

Pisanu, C., Tsermpini, E.E., Mavroidi, E., Katsila, T., Patrinos, G.P., Squassina, A., 2014. Assessment of the Pharmacogenomics Educational Environment in Southeast Europe. Public Health Genomics 17, 272, 279.

Potamias, G., Lakiotaki, K., Katsila, T., Lee, M., Topouzis, S., Cooper, D.N., et al., 2014. Deciphering next-generation pharmacogenomics: an information technology perspective. OPEN Biol. 4 (7), 140071.

Priest, V.L., Begg, E.J., Gardiner, S.J., Frampton, C.M., Gearry, R.B., Barclay, M.L., et al., 2006. Pharmacoeconomic analyses of azathioprine, methotrexate and prospective pharmacogenetic testing for the management of inflammatory bowel disease. Pharmacoeconomics 24, 767–781.

Relling, M.V., Hancock, M.L., Rivera, G.K., Sandlund, J.T., Ribeiro, R.C., Krynetski, E.Y., et al., 1999. Mercaptopurine therapy intolerance and heterozygosity at the thiopurine S-methyltransferase gene locus. J. Natl. Cancer Inst. 91, 2001–2008.

Reydon, T.A., Kampourakis, K., Patrinos, G.P., 2012. Genetics, genomics and society: the responsibilities of scientists for science communication and education. Pers. Med. 9, 633–643.

Robertson, J.A., 2001. Consent and privacy in pharmacogenetic testing. Nat. Genet. 28, 207–209.

Rothstein, M.A., Epps, P.G., 2001. Ethical and legal implications of pharmacogenomics. Nat. Rev. Genet. 2, 228–231.

Schalekamp, T., Brassé, B.P., Roijers, J.F., van Meegen, E., van der Meer, F.J., van Wijk, E.M., et al., 2007. VKORC1 and CYP2C9 genotypes and phenprocoumon anticoagulation status: interaction between both genotypes affects dose requirement. Clin. Pharmacol. Ther. 81, 185–193.

Smart, A., Martin, P., Parker, M., 2004. Tailored medicine: whom will it fit? The ethics of patient and disease stratification. Bioethics 18, 322–342.

Spear, B.B., Heath-Chiozzi, M., Huff, J., 2001. Clinical application of pharmacogenetics. Trends Mol. Med. 7, 201–204.

Squassina, A., Manchia, M., Manolopoulos, V.G., Artac, M., Lappa-Manakou, C., Karkabouna, S., et al., 2010. Realities and expectations of pharmacogenomics and personalized medicine: impact of translating genetic knowledge into clinical practice. Pharmacogenomics 11, 1149–1167.

Squassina, A., Manchia, M., Borg, J., Congiu, D., Costa, M., Georgitsi, M., et al., 2011. Evidence for association of an ACCN1 gene variant with response to lithium treatment in Sardinian patients with bipolar disorder. Pharmacogenomics 12, 1559–1569.

Stanulla, M., Schaeffeler, E., Flohr, T., Cario, G., Schrauder, A., Zimmermann, M., et al., 2005. Thiopurine methyltransferase (TPMT) genotype and early treatment response to mercaptopurine in childhood acute lymphoblastic leukemia. JAMA 293 (12), 1485–1489.

Tamaoki, M., Gushima, H., Tsutani, K., 2007. Awareness survey of parties involved in pharmacogenomics in Japan. Pharmacogenomics 8, 275–286.

Tsermpini, E.E., Assimakopoulos, K., Bartsakoulia, M., Economou, G., Papadima, E., Mitropoulos, K., et al., 2014. Individualizing clozapine and risperidone treatment for schizophrenia patients. Pharmacogenomics 15 (1), 95–110.

Vaszar, L.T., Cho, M.K., Raffin, T.A., 2003. Privacy issues in personalized medicine. Pharmacogenomics 4, 107–112.

Williams, M.S., 2007. Insurance coverage for pharmacogenomic testing in the United States. Pers. Med. 4 (4), 479–487.

CHAPTER 3

Economic Evaluation and Genomic Medicine: What Can They Learn from Each Other?

INTRODUCTION

The field of health economics is more about understanding utility and effectiveness of medical interventions rather than focusing strictly on costs. In its ideal application, health economics helps inform the decision-making process to improve the quality of health outcomes while optimizing health care expenditures.

There are three key features in health economics analyses as currently applied: (i) they are more focused on the benefits received by the health care system and society as a whole rather than the individual/patient, namely price, health state, improvements in quality of life, expansion of life expectancy, and resources saved, to name a few; (ii) the recipient of the medical intervention is, in the majority of cases, not the most informed medical decision-maker and, as such, does not have a complete picture of the potential benefits and harms of a given medical intervention or decision; and (iii) most of the time, the recipients of medical interventions do not directly pay for these treatments; rather, payment is received from a third party that the recipient supports through taxes, medical premiums, or a shared model of employer/employee contributions. As such, health economics aims to better understand the value and costs of a certain medical intervention compared with another by taking into assumption all the factors that impact on patients, health care providers, the health care system in general, and, ultimately, society.

Health economics uses various economic evaluation tools to evaluate medical interventions, including:

- Cost-effectiveness analysis (CEA): This sort of analysis aims to determine whether a medical intervention for disease diagnosis, prevention, and/or treatment improves clinical outcomes enough to justify the additional costs compared with alternative approaches. It is quantitative and relies on all the factors involved in the evaluation of medical interventions and/or health care technologies. It should be noted that CEA is not a method to show which medical intervention or health care technology reduces cost (cost—benefit analysis), but rather it is used to inform which medical interventions and/or health care technologies provide the greatest value for a given amount of health care expenditure. Another feature of CEA is that it valuates clinical events outcomes, such as cost per life-year gained, but does not allow for direct comparisons. For example, one cannot easily compare whether it is cost-effective to spend a certain amount of money to prevent a rare genetic disorder or cancer; for this, one should take into consideration other factors, such as quality of life, life expectancy, accompanying costs, and others.

- Cost—utility analysis (CUA): This approach attempts to address the aforementioned CEA limitations by measuring outcomes through a metric called a quality-adjusted life-year (QALY) that allows for comparisons across medical interventions. For example, if fewer QALYs are produced when the same amount is spent to prevent a rare genetic disorder compared with that spent to prevent cancer, this adjustment allows a better-informed societal medical decision to be made to invest in one intervention compared with the other.
- Cost-minimization analysis (CMA): This approach aims to reduce costs while keeping the outcome of a medical intervention the same. To achieve this, CMA assumes that the outcome of two different medical interventions is the same (clinical equipoise), which is rarely the case. As such, this approach cannot be commonly applied to evaluate clinical interventions.
- Cost—benefit analysis: This analysis gives a monetary value to every aspect of health care and medical intervention, which can be very challenging in health care because health care providers are often very reluctant to place a monetary value on health; as such, this approach is truly difficult to perform accurately.
- Cost-threshold analysis: This type of analysis operates in reverse in that the analyst defines a threshold of cost-effectiveness (usually derived from the calculated cost-effectiveness of a given intervention). Once this is defined the analyst asks the question, what are the necessary performance characteristics of a given test or intervention such that the cost of the test/intervention meets or exceeds the threshold.

This analytic approach is being used more frequently by developers to understand whether their test or intervention is ready for clinical use or needs additional development (Rubenstein et al., 2005; Gordon et al., 2014). In genomics this has been used to assess when a genomic risk panel or pharmacogenomics test would have sufficient discriminatory power to justify its use in clinical care (Bock et al., 2014).

All these approaches consider the cost of the medical intervention itself together with the accompanying costs, but they differ in how they measure the outcome or utility of an intervention. The two more commonly used approaches in the field are CEA and, most importantly, CUA.

GENOMICS IN HEALTH ECONOMICS

As indicated, application of economic evaluation in genomic medicine is not so different compared with other aspects of medicine. There are still some important elements that should be taken into consideration during the various stages of analysis. Such elements are mostly the type of genomic data and the various ethical, legal, and social issues that pertain to these data.

Pharmacogenomics is a core component of genomic medicine and, as such, it is used as an example to highlight the application of economic evaluation in genomic medicine. Pharmacogenomics is a new discipline in health care that attempts to enrich our understanding of how medicines work in each individual based on genomic contributions to a medicine's safety and efficacy (reviewed by Squassina et al., 2010). The latter can lead to a more efficient and effective approach to drug discovery. Furthermore, pharmacogenomics may lead to a more diversified and targeted portfolio of diagnostics and therapies, which, when used together, would yield greater health benefits to society.

The term *"pharmacogenetics"* was first used in 1959 by F. Vogel to describe a scientific discipline that sought to understand how an individual's genetic profile may influence their drug responses. Pharmacogenetics is referred to as the study of the effect of genomic variations on drug response in terms of both drug metabolism (pharmacokinetics) and drug action (pharmacodynamics). Additionally, genetic variants have been shown to explain what had previously been considered to be idiosyncratic adverse drug events (ADE). In other words, this discipline aims to identify the best medicine for a specific disease when the disease occurs in a patient population with a particular genotype. Considering the fact that genetic factors account for 20–95% of the observed responses to drug therapies, one could understand the impact of this new discipline in modern medicine. It is important to note that other factors such as age, food intake, drug–drug interactions, and the simultaneous presence of other diseases (comorbidity) influence an individual's drug response independent of, in conjunction with, or in addition to genetic factors. Pharmacogenomics is a broader term aiming to systematically assess the way genomic pathways affect disease susceptibility, pharmacological function, and drug disposition and response. It aims not only to identify genomic biomarkers for disease classification, staging, and diagnosis within the context of drug responses but also to optimize drug discovery with the goal of achieving a more desirable pharmacological response. Currently, the term "pharmacogenomics" is broadly used to cover all of these, and it impacts both new and existing medicines. It is expected to have a major impact on the translation of early-stage projects into medical treatments.

As indicated in the previous chapter, cardiology is among the medical specialties in which pharmacogenomics is used. Warfarin is the most commonly used anticoagulant, aiming to prevent and treat blood clots. Anticoagulation caused by warfarin is due to the inhibition of vitamin K epoxide reductase, an enzyme that activates vitamin K to produce anticoagulation factors II, VII, IX, and X. Warfarin is metabolized by the CYP2C9 enzyme. Cardiologists commonly prescribe it for patients with a history of atrial fibrillation, deep vein thrombosis, recurrent stroke, or pulmonary embolism, as well as in cases of heart valve replacements. A major challenge in treating patients with warfarin is that the optimal dose varies significantly from individual to individual. If the prescribed dose is too high, patients are at increased risk for serious bleeding; however, if the dose is too low, patients are at increased risk for a thrombotic event that could result in a stroke or other vaso-occlusive event. The highest risk for these complications lies within the first 30–60 days after the beginning of warfarin treatment.

Individual characteristics and behavior, such as age, sex, and diet, are some of the factors that account for the variation in warfarin dose across individuals, although these factors account for, at most, 50% of the interindividual variability (Flockhart et al., 2008). Importantly, genomic variants in the *CYP2C9* gene create variant alleles that have been found to reduce the activity of CYP2C9, thus decreasing warfarin's clearance. The variant *CYP2C9*2* allele decreases warfarin clearance by approximately 30% and the *CYP2C9*3* allele by approximately 80% when compared with the wild-type *CYP2C9*1* allele. As a result, patients with a *CYP2C9*1/*1* genotype require a daily mean maintenance warfarin dose that is higher than what is required for *CYP2C9*1/*2* heterozygote and *CYP2C9*1/*3* heterozygote patients, with the latter requiring the lowest dose. In various population groups, the variant alleles present with varying frequencies, with *CYP2C9*2* and *CYP2C9*3* being more common in European Americans and less common in Asians and African Americans, respectively. In addition, genomic variants in the *VKORC1* gene, which encodes the production of vitamin K epoxide reductase, have also been shown to affect warfarin treatment. There are five *VKORC1* variant combinations (e.g., haplotypes) that are associated with altered *VKORC1* gene expression and, as such, with different warfarin dose requirements. The allelic frequencies of these *VKORC1* haplotypes also vary in different populations. The combination of the *CYP2C9* and *VKORC1* genomic variations appears to account for another 30–40% of inter individual dose variation (Flockhart et al., 2008). Remaining unexplained variability could be due to other genetic variants, uncharacterized variants in other genomic loci, as well as other personal or environmental factors yet to be identified. Currently available warfarin dosing calculators (e.g., http://warfarindosing.org) use a combination of clinical and genetic factors and have been demonstrated to be superior in predicting the stable warfarin dose when compared with clinical judgment alone.

On average, one-third of the population carries one or both of the *CYP2C9* and *VKORC1* genomic biomarkers that are shown to be associated with slower warfarin metabolism, which in turn increases the likelihood of excess anticoagulation and the associated risk of serious bleeding. It is important to note that individuals who are "wild-type" require slightly higher warfarin doses than the recommended starting dose (6 mg/day compared with 5 mg/day). Currently, the appropriate dose is determined by regularly monitoring the anticoagulation levels through blood tests and decreasing or increasing the warfarin dose if the international normalized ratio (INR) is too high or too low, respectively. As such, pharmacogenomic testing could potentially identify the patients who are likely to present with slower warfarin metabolism, which could influence both the dose of warfarin and the recommended timing of INR studies. This may be a cost-effective way to reduce bleeding events from warfarin, as demonstrated in an economic modeling analysis based on the results of a small prospective study (Meckley et al., 2010).

According to previously published reports in 2004 and 2005, side effects from just three drugs were responsible for one-third of all emergency hospitalizations of seniors (age 65 years or older) in the United States who experienced adverse reactions to these medications. Warfarin was one of these drugs, accounting for 58,000 emergency hospitalizations per year. Also, the Adverse Event Reporting System of the US Food and Drug Administration (FDA) provides evidence that warfarin is among the top 10 drugs with the greatest number of serious adverse drug reactions. Literature reports of major bleeding frequencies for warfarin vary from as low as 0% to as high as 16%. On the basis of these data, the FDA added a new black-box warning to the warfarin label in 2006. Also, in August 2007, the US FDA updated the warfarin product label to add pharmacogenetic information. In January 2010, the agency added specific instructions regarding how to use genotype to predict individualized doses; the new label provides a concise table of dosing recommendations, stratified by genotype. However, to date, the FDA black-box warning does not require pharmacogenetic testing to be performed prior to initiation of warfarin. An evidence-based practice guideline for pharmacogenetically informed warfarin dosing has been published by the Clinical Pharmacogenomics Implementation Consortium (Caudle et al., 2014).

TRIANGULATING GENOMICS WITH HEALTH TECHNOLOGY ASSESSMENT AND HEALTH ECONOMICS

In the previous paragraph, we have briefly discussed various approaches in economic evaluation and a theoretical example in the field of pharmacogenomics, demonstrating the application of economic evaluation in genomics. The previous approaches can be fruitfully combined with health technology assessment and health economics, giving rise to comparative effectiveness research, which is closely related to the way genomic information is being used by health care professionals.

Comparative effectiveness research includes a broad range of stakeholders and stakeholder-related components, such as:
- Study design
- Stakeholder analysis (citizens/general public/patients, health care professionals, policymakers) and their prioritization
- Genomic test comparisons
- Analysis of personalized medical decision-making process. This latter component is highly relevant in the whole-genome sequencing era,

considering the fact that results may differ among different patient groups on the basis of genome characteristics, which can directly impact risk stratification as well as medical interventions.

ECONOMIC EVALUATION IN THE POSTGENOMIC ERA

Whole-genome sequencing, accompanied with competent data interpretation and genetic counseling, will revolutionize genomic medicine (Gullapalli et al., 2012; Kilpinen and Barrett, 2013). As such, its cost-effectiveness depends on the outcomes being measured, such as number of base pairs sequenced per monetary unit, the number of clinically meaningful genomic variants identified, actionable findings (both intentionally sought and incident) with their attendant clinical interventions, and (most importantly) patient outcomes. Based on these assumptions, we could consider a model to assess potential benefits and threats of whole-genome sequencing as far as its application in genomic medicine is concerned, namely the prevalence of the variants found in a certain population group, the cost of the whole-genome sequencing, the coverage and error rate, the number of incidental findings, the severity of the disease(s), and the accompanying costs and expected outcomes of the consequent medical interventions.

Overall, one could identify the following key challenges for economic evaluation in genomic medicine. First, the gradual decrease of whole-genome sequencing costs to less than $1,000 may discourage targeted genomic test development. This may, at some point, be true for developed countries, but it also may be an urgent need for developing and low-resourced countries.

Second, data interpretation and storage accompanied by the need for expert genetic counseling may create a market niche and, as such, a value proposition. As such, health care cost reimbursement entities, both public and private, in various health care systems should be adapted to accommodate this need.

Another challenge would be the development of policies to sufficiently capture personalized medical outcomes from whole-genome sequencing. Although molecular diagnostics does not necessarily increase lifespan, individuals would most likely highly appreciate this information because they can adapt their lifestyle to improve their quality of life. Because such economic evaluation approaches to capture the benefit of reporting incidental findings to patients are currently lacking, new models and tools have to be developed based on recently described analyses, such as conjoint analysis and discrete choice experiments, that model the value of knowledge/information to patients (Basu and Meltzer, 2007; Grosse et al., 2008; Regier et al., 2009). This can be particularly helpful considering the fact that whole-genome sequencing can be performed once in a lifetime (e.g., at birth) and sequencing results can be stored and ideally can be readily available for health care professionals to prescribe the proper medical interventions to patients. The latter can be cost-effective because no extra costs have to be spent for an adult patient to obtain the results of a pharmacogenomic test for individualizing his/her anticoagulation treatment because such data would already be available for him/her since birth.

Finally, the various components of next-generation sequencing involving targeted, whole-exome, and/or whole-genome sequencing should be modeled, particularly because it is not known how patients and reimbursement entities will respond to the data that will become available from genome sequencing, especially if such data impact their daily lifestyle. Such components would also include the need for further investment in genomics services, genetic counseling, data storage and retrieval, services to be reimbursed or not, incidental findings, and, ultimately, new concepts for medical treatment. The astute reader will be aware that current economic modeling approaches are not readily applicable to the multitude of interventions and outcomes that could be impacted by the availability of genomic information across the lifespan (see Chapters 7 and 9). However, it may not be necessary to model all relevant scenarios because certain common scenarios that would use genomic information could justify the cost of the once-in-a-lifetime sequencing approach, meaning that all the other genomic information could be considered to be value added without incremental cost. This assumes that the information would be used appropriately and would not result in either harm to patients or waste of resources, which is hardly a robust assumption in the current health care delivery environment. Nonetheless, this approach may be feasible, but it has yet to be accomplished.

REFERENCES

Basu, A., Meltzer, D., 2007. Value of information on preference heterogeneity and individualized care. Med. Decis. Making 27 (2), 112–127.
Bock, J.A., Fairley, K.J., Smith, R.E., Maeng, D.D., Pitcavage, J.M., Inverso, N.A., et al., 2014. Cost-effectiveness of *IL28* genotype-guided protease inhibitor triple therapy

versus standard of care treatment in patients with hepatitis C genotypes 2 or 3 infection. Public Health Genomics 17, 306–319.

Caudle, K.E., Klein, T.E., Hoffman, J.M., Muller, D.J., Whirl-Carrillo, M., Gong, L., et al., 2014. Incorporation of pharmacogenomics into routine clinical practice: the clinical pharmacogenetics implementation consortium (CPIC) guideline development process. Curr. Drug Metab. 15 (2), 209–217.

Flockhart, D.A., O'Kane, D., Williams, M.S., Watson, M.S., 2008. Pharmacogenetic testing of CYP2C9 and VKORC1 alleles for warfarin. Genet. Med. 10, 139–150.

Gordon, L.G., Mayne, G.C., Hirst, N.G., Bright, T., Whiteman, D.C., 2014. Australian cancer study clinical follow-up study, Watson DI. Cost-effectiveness of endoscopic surveillance of non-dysplastic Barrett's esophagus. Gastrointest. Endosc. 79 (2), 242–256.e6.

Grosse, S.D., Wordsworth, S., Payne, K., 2008. Economic methods for valuing the outcomes of genetic testing: beyond cost-effectiveness analysis. Genet. Med. 10 (9), 648–654.

Gullapalli, R.R., Lyons-Weiler, M., Petrosko, P., Dhir, R., Becich, M.J., LaFramboise, W.A., 2012. Clinical integration of next-generation sequencing technology. Clin. Lab. Med. 32 (4), 585–599.

Kilpinen, H., Barrett, J.C., 2013. How next-generation sequencing is transforming complex disease genetics. Trends. Genet. 29 (1), 23–30.

Meckley, L.M., Gudgeon, J.M., Anderson, J.L., Williams, M.S., Veenstra, D.L.A., 2010. Policy model to evaluate the benefits, risks and costs of Warfarin pharmacogenomic testing. Pharmacoeconomics 28 (1), 61–74.

Regier, D.A., Friedman, J.M., Makela, N., Ryan, M., Marra, C.A., 2009. Valuing the benefit of diagnostic testing for genetic causes of idiopathic developmental disability: willingness to pay from families of affected children. Clin. Genet. 75 (6), 514–521.

Rubenstein, J.H., Vakil, N., Inadomi, J.M., 2005. The cost-effectiveness of biomarkers for predicting the development of oesophageal adenocarcinoma. Aliment. Pharmacol. Ther. 22 (2), 135–146.

Squassina, A., Manchia, M., Manolopoulos, V.G., Artac, M., Lappa-Manakou, C., Karkabouna, S., et al., 2010. Realities and expectations of pharmacogenomics and personalized medicine: impact of translating genetic knowledge into clinical practice. Pharmacogenomics 11 (8), 1149–1167.

Vogel, F., 1959. Moderne probleme der humangenetik. Ergebn. Inn. Med. Kinderheilkd. 12, 52–125.

CHAPTER 4

Introduction to the Technical Issues of Economic Evaluation

INTRODUCTION

Cost-effectiveness analysis (CEA) is defined as an analytical technique intended for the systematic comparative evaluation of the overall cost and benefit generated by alternative therapeutic interventions for the management of a disease (WHO Guide to Cost-effectiveness Analysis, 2003). The cost associated with the different treatments is measured in monetary units, whereas the benefit from the interventions is estimated in similar physical units that can be determined objectively by suitable measurements. Therapeutic interventions usually diminish the impact of the disease, eliminate its symptoms, improve quality of life, and also prolong survival when this is possible. However, society's health care needs increase over time; the associated costs are too great and the available resources are too limited to meet current needs.

The idea of comparative evaluation of alternative interventions is fundamental in economics and is related to the way in which economists assess the value of the relevant options. Economic evaluation, a major part of which is CEA, is a standard approach to the problem of evaluating alternative interventions through their opportunity cost. This term is used to describe the process of evaluating different health technologies by considering the opportunity cost, meaning the best possible alternative options at the societal and the individual levels (Palmer and Raftery, 1999). This cost should not be confused with the explicit cost of acquiring a technology, which is constant and determined based on historical information; instead, it should be evaluated periodically because it is a function of current practices and changes over time as new treatments become widely available to the general population.

The application of CEA allows such a comparison to be made, and the limited resources that the health systems can allocate can be used effectively based on optimization behaviors grounded in the tenets of economics (Canning, 2009). As already mentioned, it is still (at a theoretical level) a basic tool for the rational distribution and a central point of reference for health economics methodology. This type of analysis serves three primary goals: determination of the price of a technology; definition of the level of insurance compensation; and drafting of guidelines to be used as guides for health care professionals when prescribing.

UNIQUE ASPECTS AND DIFFERENTIATION OF CEA FROM ALTERNATIVE APPROACHES

First, this method focuses on evaluating an intervention in physical units but is more interested in determining life expectancy (Ramsey et al., 2005) and/or quality of life (in which case it is called "cost–utility analysis"; Torrance, 1986; O'Brien and Gafni, 1996); this means that the method deals primarily with the patients' ultimate *health outcomes* instead of intermediate indicators. Such intermediate/clinical indicators are of interest only to the clinical scientist when making a diagnosis, proposing treatment, or deciding on future research, but are not particularly useful for economic evaluation except in the few circumstances when an intermediate indicator is associated with a health outcome through a robust chain of evidence.

As an example, the benefit achieved by administering insulin to a patient with diabetes could be defined in terms of reducing the levels of glycosylated hemoglobin; other such indicators could be the reduction of hypertension or low-density lipoprotein (LDL) cholesterol in cardiac patients. All of these indicators are aimed at reducing a risk factor and, thus, at leading to increased survival or better quality of life. Instead, economic evaluation deals with clinical intermediate indicators only to the extent necessary to convert them into ultimate health outcomes (usually through statistical models).

Another example is the measurement of the progression-free survival (PFS) in oncology. This is of particular clinical interest because it defines the period over which the patient is free from disease progression, but it is not used indiscriminately in economic evaluation because it is often unable to demonstrate the physical equivalent of clinical observation, such as prolongation of life or improved life expectancy (Sullivan et al., 2011).

In this sense, economic evaluation "ignores" a drug's mode of action, its route of administration, or its pharmacokinetic properties. Therefore, provided a certain technology has been approved by the competent authorities for use in the general population, it is assessed by the single indicator that is of the most interest for insurance carriers and for individuals in charge of health care budgets: patient survival (and the relevant quality of life). With this "flexible" (but objective) approach, it is possible to calculate the comparative cost and benefit generated by completely different health technologies targeting the same disease, as long as the outcomes are measured on the same scale (months or years of life) and the cost is presented in a common unit of measurement (€ or $).

At the same time, economic evaluation is distinct from biostatistical analysis, which determines the (clinically oriented) difference in survival between alternative interventions. In this case, the indicator used when dealing with patient data is the median survival (Collett, 2003) for each sample; the statistical significance of their difference is estimated, usually by use of survival curves, whereas economic evaluation deals exclusively with mean survival (National Institute for Health and Clinical Excellence, 2011). The use of median survival is justified in statistical analysis because mean survival is often falsely elevated. This happens because some patients will live for a long time after responding to a treatment, thus increasing mean survival, whereas median survival is quite lower. In such cases of right-skewed samples (Gray et al., 2011), it would be incorrect to use the mean because this would be overestimated and, therefore, so would the relevant intervention's effectiveness (and, accordingly, its cost).

In economic analysis, the mean is used instead because it is the most suitable measure for calculating the actual economic effects of the adoption of a new technology by society (Barber and Thompson, 1998). In this sense, the mean does not accurately reflect, from a clinical standpoint, the differentiation between interventions, nor does it examine how the life–years gained are "shared" among the patients receiving treatment in a statistical sample or in the population. However, one of the basic statistical properties of the mean is that it equalizes the distances (differences) between the highest and lowest values, thus permitting an approximation of the true overall benefit or cost of an intervention, whereas the median as a robust estimator is not affected by outliers as a measure of central tendency. The problems of asymmetry and of inductive analysis with inclusion of the mean are addressed with the use of recursive simulation techniques (bootstrapping when involving patient data or Monte Carlo for distributions; Efron and Tibshirani, 1993). Extensive discussions of these methods are beyond the scope of this book.

AN APPROPRIATE MEASURE FOR CEA

In the technical application of the method, the appropriate measure is the incremental benefit obtained by a therapeutic intervention (usually an innovative treatment) and the associated cost in comparison with the treatment that is considered to be the gold standard for that particular disease, and the ratio of their differences (Drummond et al., 2005). The concept of the marginal quantity, which is used in classical economics to perform analyses associated with alternative options, is not used in health economics for practical reasons because health technologies do not produce results infinitely divisible at the physical level. The mathematical concept of the derivative for infinitesimal changes, which is a feature of marginal quantities, can be adequately replaced by incremental quantities without loss of generality for method application.

The size of this difference ratio, mentioned previously, is the major criterion for rejecting or accepting a new treatment. The difference in average cost-effectiveness ratios (ACERs) for the alternative treatments may appear to be a consistent indicator, but it does not describe the true benefits of the new technology (Detsky and Naglie, 1990). In a usual case, the mean cost of the cheaper and older technology is lower compared with that of a new innovative treatment, but its administration is associated with reduced survival for the patient and, therefore, lower societal utility. The patients, their families, and the organizations managing public health look to technological progress and the attendant life prolongation, and so they need to know the additional amount that must be invested to adopt the new technology. This incremental cost in comparison with the benefit from the new technology (which is the appropriate measure) partly reflects the rate of return on the industry's R&D investment, the reward for innovation, and society's preference for the adoption of new technologies under financially viable terms. Therefore, and despite the fact that the new technology is associated with greater mean costs, as expected, it is also associated with greater benefit as the society requires.

Another issue arises when the mean cost is used for the comparison of mutually exclusive alternative interventions. In some cases, the option of nonintervention (the "do-nothing" option) for the treatment of a disease is associated with benefits at no cost. In such a case, the calculation of the mean cost will arbitrarily alter the result and the calculated indicator will be misleading, as shown in Figure 4.1.

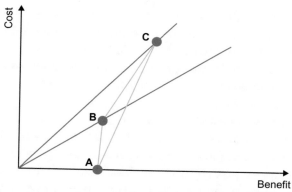

Figure 4.1 Mean versus incremental cost and the "do nothing" option.

According to the diagram, the mean cost associated with option B is lower than option C (because they lie on lines with different slopes); therefore, it should be "preferred" over the latter based on the simplistic criterion of mean cost per unit of benefit. Nevertheless, a society that wishes to improve the level of health for its citizens might choose point C, which is associated with greater benefit (survival). Additionally, point A has zero cost because nonintervention is associated with a small benefit and, by this reasoning, it should be preferred over the other two. In reality, society will choose between option A and option C or a linear combination of the two if the two technologies were theoretically fully divisible, considering that the border of the technologies' productive capacities is defined by the line segment connecting these two alternatives. Option B can never constitute optimizing behavior, as already shown (Cantor, 1994), because it lies within the convex polygon defined by the origin, option A and option C. In such a case, a true decision criterion would be the slope of the line defined by these points A and C. If this slope (threshold), which indicates the additional cost of an additional year of life, were attractive based on society's priorities and strategic choices, then the new technology would be adopted; otherwise, we would be led to the option of nonintervention (point A).

CEA DECISION CRITERIA

Let us consider two different treatments, T (new treatment) and S (standard treatment), each associated with a specific effectiveness (E) and cost (C) for the management of a disease. E_T, E_S, C_T, and C_S correspond to the mean effectiveness of the new treatment, the mean benefit of the standard treatment, the mean cost of the new treatment, and the mean cost of the standard treatment, whereas $\Delta E = E_T - E_S$ and $\Delta C = C_T - C_S$ are the definitions of the differences in cost and effectiveness, respectively.

Comparative evaluation will give the following four possible scenarios (Drummond et al., 2005):

A. $\Delta E = E_T - E_S > 0$ and $\Delta C = C_T - C_S > 0$

This is the most common case. Usually, an increase in mean survival with the innovative treatment and a corresponding increase in overall cost associated with its administration are observed. In recent years, new discoveries have been made continuously as new drugs, new biological and gene therapies, new diagnostic methods, targeted treatments, or new procedures have been developed. Technology, in most cases, significantly increases the cost of services because of the great developing, purchasing, and operating expenses; the price of technology (which is a fraction of the overall cost of the intervention) tends to incorporate these costly processes. In such cases, there should be a criterion by which to assess whether the increased cost is justified by the additional effectiveness (Figure 4.2, Box A).

B. $\Delta E = E_T - E_S > 0$ and $\Delta C = C_T - C_S < 0$

In this scenario, the new treatment dominates because it provides more effectiveness and is associated with lower cost than the standard treatment. In this case, the previous technology must be completely discarded and the new treatment must be fully adopted by the health system (Figure 4.2, Box B).

Medical and/or service outcomes	Cost of care decreased	Cost of care unchanged	Cost of care increased
Improved	B		A
Unchanged			
Worsened	D		C

Figure 4.2 The value framework, adjusted. Boxes A–D depict various scenarios (see text for details).

C. $\Delta E = E_T - E_S < 0$ and $\Delta C = C_T - C_S > 0$

This unpromising scenario for the new technology includes increased cost but lower effectiveness from its use compared with the standard treatment. According to the applicable terminology, the new treatment is dominated by the older one and should be discarded because it is not in the society's best interest and is not a good health investment (Figure 4.2, Box C).

D. $\Delta E = E_T - E_S < 0$ and $\Delta C = C_T - C_S < 0$

In this fourth scenario, the difference in survival is for the standard treatment, even though the use of the new treatment is associated with resource savings. In scenario D (as in scenario A), the ultimate decision to adopt or reject the new technology will be based on weighing the savings (which is the desirable outcome) and the reduced effectiveness (which is a negative consequence) when comparing treatments. This criterion is essentially the same as in the first scenario, but the level of the indicator may differ (Figure 4.2, Box D).

A special case of these scenarios is the comparison of costs only when clinical practice has failed to show a difference in favor of one of the compared technologies. In this case, the treatment associated with the lowest overall cost will be selected. This type of analysis is called cost-minimization analysis and is not considered a complete economic analysis because it does not examine the relative benefit of the treatments. Nevertheless, we should note that failure to demonstrate a statistically significant difference between treatments in a study does not necessarily mean that the interventions are equivalent. "Statistical significance" is a widely used concept but is based on an arbitrary level of significance determined from the literature. Economic evaluation, however, is interested in the mean estimate of quantities (i.e., the mean expected difference between interventions) (Claxton, 1999). Let us imagine a very simplistic example. Assume that we live in a world without costs; in which case, the only thing we would want to compare between two therapeutic interventions would be their effectiveness measured in "years of life gained." As in standard statistical practice, if the confidence interval of the difference in survival between interventions includes zero, then we may say that there is no statistically significant difference between treatments, at a possibility of 95%, for the population from which the samples were taken. In practice, however, we are not interested in inductive conclusions; rather, we need to decide whether to administer one treatment or the other, disjunctively (i.e., with no intermediate alternative). In such a case, we would select the treatment with the greatest expected mean value, regardless of the confidence interval. Therefore, it is possible to perform CEA even without a statistically significant difference in survival between interventions (Briggs and O'Brien, 2001).

If, however, the equivalence of the interventions is assured, then comparing costs will reveal the maximizing option and will lead to the adoption of the least expensive technology. In the marginal case where neither cost nor benefit differs, although the patients' quality of life is—or may be reasonably assumed to be—equal, the economic analysis is unable to provide an answer and other clinical and societal criteria must be considered to reach a decision. The cases described here are presented graphically in Figure 4.3 (Black, 1990).

The axes show the difference between cost and benefit for interventions T and S; the differences are indicated with the Greek *delta* (Δ). Quadrant I represents the first scenario. The new technology increases survival as well as cost compared with the standard treatment. If the additional cost is not forbidding, then the new technology will be adopted; however, if the cost is considered excessive, then it will be rejected. There is also a "gray zone" where the additional benefit is associated with higher cost and the result will be uncertain until the expense considered

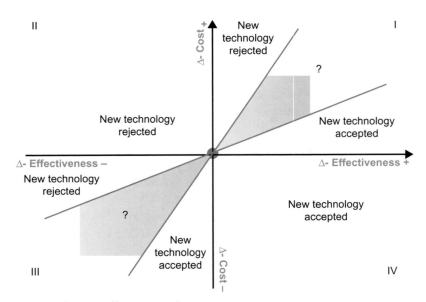

Figure 4.3 The cost-effectiveness diagram.

acceptable for an additional year of life is fully quantified. Scenario B is represented in quadrant IV, where T is cheaper and better. No additional criteria are needed here; the new treatment should be selected immediately and reimbursement for S should be discontinued. The same logic is followed for scenario C, where T is more expensive and less beneficial; therefore, T should be rejected. Quadrant III represents scenario D, where the resources saved by treatment T are associated with reduced survival. In this case, too, we would need a fully defined quantitative criterion to compare treatments. If there is no difference between treatment benefits, then we move along the vertical cost axis and we compare only the costs of the interventions. If T is more expensive, then we are in the upper area of quadrant I and in quadrant II, in which case the treatment is rejected; otherwise, we are located in the lower part of quadrant IV and in quadrant III, in which case we accept the treatment based on the cost-minimization analysis mentioned previously.

Note that according to the diagram, the angles of the slopes (thresholds) for scenario A and scenario D are identical, indicating that the additional investment we are willing to make to obtain an additional year of life is equal to the savings we will demand for losing 1 year of life. However, as a society, we are continuously less willing to lose 1 year of life compared with standard treatment (Morrison, 1998), and usually the amount of savings required is quite larger than the amount we would be willing to pay to obtain an additional year of life; it can be as high as twice the first in the case of health services (O'Brien et al., 2002). This case is represented in Figure 4.4.

Figure 4.4 reveals that the area of rejection of the new technology is now greater in quadrant III because the indifference angle is smaller than in quadrant I. The mathematical formula that quantifies the ratio of differences is called *incremental cost-effectiveness ratio* (ICER) and is described as follows:

$$\text{ICER} = \Delta C / \Delta E \text{ or ICER} = (C_T - C_S)/(E_T - E_S)$$

By rearranging this formula, we find that:

$$\text{ICER} = \frac{1}{\Delta E} \Delta C = \text{NNT} \times \Delta C$$

The number needed to treat (NNT) symbol is known from epidemiology as the number of patients who need to follow the new treatment for an additional year of life to be obtained (http://www.cebm.net/index.aspx?o=1044). By this logic, ICER is the product of the NNT multiplied by the individual incremental cost.

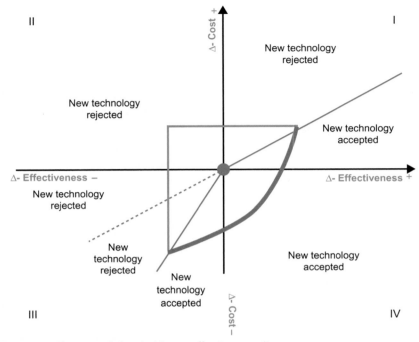

Figure 4.4 The curved-threshold cost-effectiveness diagram.

The ratio of average cost differences (ACER), which we mentioned previously, is calculated here, and it is obviously not equal to the ICER in the general case, except for specific values that make them equal to each other.

$$\text{ACER} = (C_T/E_T) - (C_S/E_S) \neq (C_T - C_S)/(E_T - E_S) \text{ or ACER} \neq \text{ICER}$$

The amount a society is willing to pay to obtain 1 year of life is called *willingness to pay* (WTP) and is represented by the Greek letter *lambda* (λ) (Pauly, 1995), whereas the *willingness to accept* (WTA) expresses the savings a society demands for losing 1 year of life, as mentioned previously. The λ is the slope of the line dividing the cost-effectiveness diagram in two. If the ICER is more than λ, then the new technology will be rejected; however, if it is smaller than the new technology, then it is a socially beneficial option because it is cost-effective. If the WTP is greater than the ICER, then the formula can be rearranged as follows:

$$\lambda > \text{ICER} \acute{\eta} \lambda > \frac{\Delta C}{\Delta E} \ \acute{\eta} \lambda \chi \Delta E > \Delta C \text{ or } \lambda \chi \Delta E - \Delta C > 0$$

The increase in benefit is greater than the increase in cost for points within the first quadrant, given that λ is by definition a positive number, whereas if $\Delta E < 0$ and $\Delta C < 0$, then the term $\lambda \chi \Delta E$ is a negative number and the term $-\Delta C$ is a greater positive number; therefore, the cost (savings) is greater than the benefit (obtained by S) and treatment T is again a cost-effective option.

In reality, ΔE and ΔC are not simply two data points (values); they are actually distributions with an estimated mean and a standard deviation (just like the ICER). Consequently, in some cases it is convenient to alter the ICER equation with a suitable linear transformation. Because the ICER is an index (ratio) of two other statistical measures (ΔC and ΔE), it does not follow any known statistical distribution; instead, it presents certain assessment difficulties when the difference between treatments is small. In this case, the limits become infinite, which complicates the statistical manipulation of the results. If S has a marginally greater benefit, then the limit of ICER tends to negative infinity; however, if T has slightly greater survival, then the ICER tends to positive infinity. Let us assume, for example, a value of $\Delta C = €10,000$ and of $\Delta E = 0.001$ or $\Delta E = -0.001$. At these values, the ICER obviously tends to infinity. The ICER presents one additional difficulty: it does not provide clear information about which of the treatments being compared should be adopted if we do not already know the quadrant we are in, such as if the signs of ΔE and ΔC are unknown. In this case, problems in estimating 95% confidence intervals would arise if we were to classify negative ICER ratios together with positive ones.

To overcome such difficulties in difference ratios, the following transformation has been proposed:

$$\lambda \chi \Delta E - \Delta C = \text{NMB}$$

where NMB is the *net monetary benefit* arising from the difference in benefit and in cost for the treatments being compared and for a given value of λ. A therapeutic option in this case is considered acceptable if (and only if) NMB > 0, regardless of the sign of ΔE.

A diagrammatic description for NMB and λ compared with the ICER is given in the cost-effectiveness diagram (Figure 4.5, where b_λ is the NMB).

The diagram shows that the slope of the WTP (i.e., the λ) is greater than the slope of the ICER, which proves that option A is more

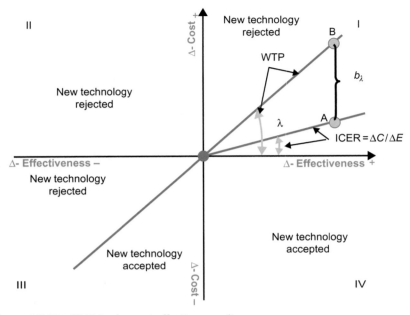

Figure 4.5 The NMB in the cost-effectiveness diagram.

advantageous for the society. The society would be willing to marginally accept up to option B, which is associated with greater cost than A for equal intervention effectiveness. The vertical difference between the two points expresses the net incremental benefit from the new technology. As a result, the most significant advantage of NMB is the fact that it is expressed in linear form, it is drawn as a straight line on inductive analysis, and its statistical manipulation is easier because of its linearity and the fact that it can be performed using the analytical approach of classical statistical distributions to calculate 95% confidence intervals (Willan and Lin, 2001).

The variance (V) of the NMB can be defined with the following equation:

$$V(\text{NMB}) = \lambda^2 V(\Delta E) + V(\Delta C) - 2\lambda \text{COV}(\Delta E, \Delta C)$$

and the 95% confidence interval can be defined as:

$$(\text{NMB}) \pm z_{(a/2)} \sqrt{\sigma^2_{\text{NMB}}}$$

where z represents the normal distribution, σ^2 represents the variance, and σ represents the standard deviation. The same concept of linear transformation can also be applied to the health benefit. In this case, all quantities are converted to health care terms (terms of benefit, not cost as before). This approach is equivalent to the one we presented, but the quantity we estimate is called NHB (*net health benefit*). Generally speaking, the NBM is superior to the ICER, with the only difficulty being that the calculations of variance and confidence intervals require that the λ be known beforehand.

THE CONCEPT OF EXTENDED DOMINANCE

Based on what we have seen so far, if a treatment is more effective and less expensive compared with a competitive treatment, then it is said to dominate. However, we should examine another type of dominance that occurs when comparing three or more options.

Let us start at the beginning. Usually, innovative health technologies with much greater costs are developed over time. This case is depicted in Figure 4.6.

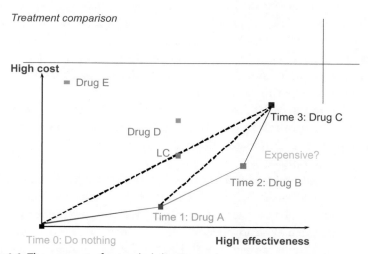

Figure 4.6 The concept of extended dominance.

It should be noted that the new health technologies (drug A, drug B, and drug C) are associated with much greater costs when they provide greater effectiveness. Note that the transition from drug B to drug C, for example, which achieves little additional effectiveness, requires the sacrifice of a large amount of resources (large additional cost); this is not the case for drug A and drug B, and even less so for "do nothing" and drug A. This situation is quite common today as new technologies are developed, and this puts into question the future survival of health systems. To achieve a small additional benefit, very large amounts of resources need to be sacrificed. Technically, in this case the ICER between option "do nothing" and drug A is smaller than the ICER between drug A and drug B, which is smaller than the ICER between drug B and drug C. The condition of extended dominance occurs when an option is dominated not by an alternative option, as seen in a previous chapter, but by a linear combination of two other options. For example, drug D is dominated by the linear combination of "do nothing" and drug C at linear combination and is therefore "extendedly dominated" (the same thing is true for the linear combination of A and C). Visually, it lies to the *left* of the line connecting options "do nothing" and "drug C," whereas its effectiveness is *between* those two options. In such a case, we could achieve the same benefit as that achieved with drug D, but by expending much fewer resources if we were to use a combination of technologies "do nothing" and "drug C." Note that this does not occur for any other drug between A, B, and C. Visually, when any new technology lies to the right of any linear combination of the other possible options, there is no extended dominance. If it lies to the left (and "between" them with regard to effectiveness), then there is extended dominance. So a simple way to check for this situation when comparing options is to make a diagrammatic representation of our options. If this situation does exist, then all extendedly dominated (D) or dominated (E) options should not be adopted by the health system. E is dominated because it is the least effective and most costly of all the options.

Another analytical method is to record all the available options (e.g., in an Excel spreadsheet) by order of increasing effectiveness and indicate the associated cost for each one. If the transition to a more effective technology is associated with lower cost, then we have a case of dominance. After excluding these options, the ICER for the remaining options is calculated. If each ICER is greater than the next, then we have a case of extended dominance and those options should also be disregarded.

An example of the calculation of dominance and extended dominance over five available options is given. Assume the following options:

Step 1: Rank order of all options by increasing benefit.

Option	Benefit (years)	Cost (€)
A	12	12,500
B	13	13,000
C	11	23,000
D	10	24,000
E	8	8,000

Result:

Option	Benefit (years)	Cost (€)
E	8	8,000
D	10	24,000
C	11	23,000
A	12	12,500
B	13	13,000

Step 2: Remove all directly dominated options.
Result:

Option D is directly dominated by option C because it offers less benefit at greater cost. Option C is also directly dominated by option A for the same reason. Therefore:

Option	Benefit (years)	Cost (€)
E	8	8,000
A	12	12,500
B	13	13,000

Step 3: Calculate the ICER values.
Result:

Option	Benefit (years)	Cost (€)	
E	8	8,000	1,000[a]
A	12	12,500	1,125
B	13	13,000	500

[a]This ICER is calculated in comparison to the "do nothing" option, the cost and benefit of which are set to zero.

Step 4: Check if each ICER is smaller than the next; otherwise, remove the extendedly dominant options.

Result:

Option A has an ICER greater than the next option; therefore, it is dominated by a linear combination of E and B. For example, (50% × E + 50% × B) gives (8/2 + 13/2) QALYs (= 10.5 QALYs) at a cost of €8,000 × 50% + €13,000 × 50% (= €10,500), whereas the "purchase" of 10.5 QALYs with option A requires a cost of (10.5/12) × €12,500 = €10,937 and is therefore more expensive, as expected. For a more realistic version of the example, let us imagine that we were to give treatment B to 50% of our patients and treatment E to the remaining 50% of patients.

The available options would then be as follows:

Option	Benefit (years)	Cost (€)
E	8	8,000
B	13	13,000

In conclusion, the ultimate choice between options E and B depends on the λ, that is, on whether society is willing to pay an additional €5,000 for 5 life-years per patient. It should be noted, however, that although extended dominance is an attractive concept in economics because it allows us to determine how to achieve the same benefit with the lowest possible cost, it also has a few disadvantages. For example, to achieve the same benefit with a linear combination of technologies, we need to use a health technology with a demonstrably higher benefit (B) for a certain percentage of patients and a less effective (but cheaper) technology (E) for another segment of patients. This raises the ethical issue of equal patient treatment and does not provide an accurate criterion for identifying the patients who are to receive one or the other treatment. The patients themselves, who care primarily for their own welfare and not at all for financial models, will try in every case to be included in the patient segment that receives treatment B. In addition, the owner of technology A, who will obtain demonstrably higher benefit compared with option E, will not readily resign themselves to being excluded from the market based on the abstract concept of "linear option combinations" and will attempt by any legal means to be included in the state reimbursement lists and maintain their market share.

SELECTION OF ALTERNATIVE OPTIONS DURING ECONOMIC EVALUATION

Selecting one of the alternative options under comparison is important for applied research. Such a comparison may be strictly limited to include only a new technology and the technology that was its main competitor until then. A comparison may also be made against no treatment or against the best supportive care. If justified by the type of study, then the comparison may also include the least expensive treatment (e.g., a generic drug) or the most effective treatment in terms of survival. It is obvious that an alternative should be selected with consideration of the practices followed in each individual country (Drummond et al., 2005).

VALUING THE QUALITY OF LIFE–COST–UTILITY ANALYSIS

Cost–utility analysis uses various indices and tools to measure the quality of the patient's life to adjust the result according to patient quality of life. A common measure of effectiveness is the "quality-adjusted life-year" (QALY). Quality is often measured on a scale of $0-1$, or of $0-100$, where 0 is the "worst possible" and 100 is the "highest or best possible" state of health. A "QALY" is a period of 1 year weighted by the quality of life that the patient is experiencing when suffering from a disease or when improving as a result of a treatment. For example, if a prostate cancer patient is found to have 75% quality of life, then 1 year of life with prostate cancer is equivalent to 0.75 years of life with perfect health (0.75 QALYs). If the patient improves to 90% after treatment, then 1 year of life after treatment is equivalent to 0.9 years of life in perfect health, and the treatment benefit is 0.15 years of life.

Various methodological tools are used to value a patient's health state and quality of life. Some of these are specialized for specific diseases, whereas others seek to evaluate a patient's general state of health. Some are based on simple indices and others are more comprehensive but also more difficult to assess. The subjects in such studies are usually patients, but they may also be health professionals, such as nurses or physicians, or the general population. Quality assessment may be done directly or indirectly through the use of certain characteristics of the treatment groups and the creation of empirical utility functions by professional investigators. Examples of such efforts are the EuroQol EQ-5D (http://www.euroqol.org/eq-5d-references/reference-search.html), the Health Utility

Index (HUI; http://fhs.mcmaster.ca/hug), the Quality of Welfare Scale (QWB), and the SF-36 (http://www.sf-36.org/tools/sf36.shtml). Because of the importance of the quality of life, and because this type of analysis will (in theory) facilitate broad comparisons between different medical interventions by reducing them all to a common measure of value (the QALY), cost–utility analyses are becoming more and more common, and many organizations such as the UK National Institute of Health and Care Excellence (NICE) encourage their use. A simple example of an ICER calculation in such analyses is given.

	Survival (years)	Quality of life	QALYs	Cost (€)
Treatment A	5	50%	550% = 2.5	10,000
Treatment B	4	60%	4 × 60% = 2.4	11,000

$$ICER = (11,000 - 10,000)/(2.5 - 2.4) = €10,000 \text{ per QALY}$$

This example should make it clear that 1 QALY is equal to 1 calendar year only for people in perfect health. Let us repeat that 1 QALY is "1 year of life for a person in perfect health" or, more accurately, "the equivalent of 1 year of life in perfect health."

We consider the methods used for quality assessment. The careful reader will certainly note that such assessments are not free of disadvantages and are heavily subjective. The latter should not surprise us because economics, by its nature, is the study of the "production of useful commodities" (health services in our case) under specific conditions of "preference" by the "consumers" (patients).

The simplest method of utility assessment is the visual analog scale (Figure 4.7). In this case, we ask the patients to evaluate their health on a 100-point scale like the one in this chapter. The advantage of this method

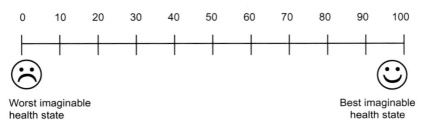

Figure 4.7 The visual analog scale.

is its calculation simplicity because it allows even a relatively inexperienced investigator to assess the patient's quality of life. Its disadvantage is the fact that it does not offer the subject the option to "trade" states of health (no questions like: "Would you prefer state X or state Y?") and therefore tends to make the subject give more pessimistic answers.

The second type of analysis is the typical game (Figure 4.8). In this case, we ask the patients to take part in a "game." We ask whether they would prefer their current state of health (option 1) or to participate in a game in which they would have P chances of being completely cured and $(1 - P)$ chances of dying (option 2). When we determine the possibility P, which makes the patients indifferent to the two options (their current state versus the game), then essentially the relevant quality of life has been assessed. For example, if patients become indifferent to whether they continue living with their specific condition or they are given a 90% possibility of becoming perfectly healthy and a 10% possibility of dying, then their quality of life is determined to be 90%.

Another method of quality assessment is the time trade-off method. In this case, we ask the patients how many years of their life they would be willing to sacrifice to become completely healthy. For example, if they are indifferent to whether they live with their condition for 10 years or live without the condition for 6 years (i.e., they would be willing to sacrifice 4 years to regain their health), then their quality of life is estimated to be 60%. This is schematically represented (Figure 4.9). Although this approach is simple and intuitive, in certain cases (e.g., rheumatoid arthritis) many patients may not be willing to trade their condition for years of life;

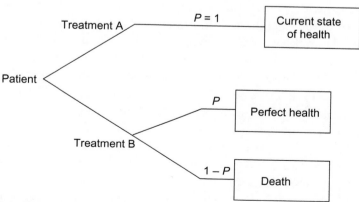

Figure 4.8 The typical game.

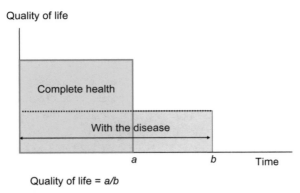

Quality of life = a/b

Figure 4.9 The time trade-off.

therefore, a different method of quality of life assessment may be necessary with the use of a different instrument or another "anchor point" instead of death.

Finally, another interesting method is quality assessment through a questionnaire. We examine the generic EQ-5D questionnaire, but the others follow the same approach. With the EQ-5D, we ask the subjects questions involving five dimensions of their health (hence "5D"). These are mobility, self-care, daily activities, pain, and anxiety. The questions are answered using a simple three-level scale of "no problems," "some problems," and "severe problems." The full questionnaire is available online at www.euroqol.org.

It should be obvious to the reader that certain answers are meaningless or require further investigation; such an example would be a patient who is bedridden with intense pain but expresses feeling no anxiety. The scoring system of the English version of the questionnaire is described. There are no scores available for some countries and so the investigator can select to use another country as a basis, either the United Kingdom or some country that they believe (or have ascertained) has prevailing preferences that are not very different from those of the country for which there are no scores available. One can easily search the Internet for the scoring algorithm of similar instruments (as well as of the EQ-5D) in Excel spreadsheet format. Starting from perfect health (1), one begins to automatically "lose points" for answering "2" to any question (plus the initial loss of -0.081; for further details go to http://www.euroqol.org/about-eq-5d/valuation-of-eq-5d/eq-5d-5l-value-sets.html). The number of additional points lost is given next to each question. If the patient replies "3" (severe problems) even to only one question, then that patient loses an additional -0.269 points. The final estimate will be the patient's quality of life (which will be a number between 0 and 1). We should note that because such analyses are based on statistical methods, the relevant scores will incorporate the flaws of these methods. For example, there may be cases with scores less than zero, cases where the loss of welfare due to the disease is nonlinear, and other cases. A detailed presentation of the advantages and disadvantages of such instruments is beyond the scope of this work.

In conclusion, if one wants to estimate QALYs using a clinical questionnaire (e.g., PASI for psoriasis or DAS28 for rheumatoid arthritis), then statistical methods should be used to convert clinical estimates to QALYs based on the literature. To do this, one would enter the clinical estimates into an equation that includes utility and therefore calculate the QALYs (e.g., an equation like $QALY = a - b \times PASI$, where a and b are calculated for specific samples through statistical regression).

A CONCEPTUAL APPROACH TO QALY

We should examine the QALY a little more, because this concept permits, at least at a theoretical level, extensive comparisons to be made between interventions originating from very different fields of clinical research. For convenience, we might imagine that the QALY has some of the same properties as money. For example, money can be used as a means of valuation, because every product or service is measured in monetary units. By this rationale, we can compare commodities by value. The QALY has the same function (in its most ambitious form). A therapeutic procedure in cardiology, a surgical operation, the treatment of oncology patients, the treatment of patients with rheumatoid arthritis, and others are actually completely different activities—in fact, they are essentially separate minor scientific fields. However, the theoretical concept of the QALY allows us to compare them and check which activity is the most "rewarding" in QALYs, that is, it offers the most QALYs at the lowest "price." This idea lays the ground for the concept of allocative efficiency. One basic question in economics is, "where should the available resources be invested to obtain the best possible result?" and its answer is the answer to the important issue of allocative efficiency. A reasonable approach would be to invest more resources into "productive activities" (conditions from which many QALYs can be obtained) and less money into the others if we were aiming at some kind of maximization of the total QALYs with no regard to who will benefit from these. A different approach would be to think less "socially" and to invest money into areas where fewer QALYs are produced, because that is probably where the "difficult" conditions belong (scourges, e.g., aggressive forms of cancer, renal failure patients, etc.) and should therefore be given priority. Yet another approach would be to somehow distribute the money "horizontally" to all conditions (e.g., by disease prevalence), thus giving all patients a chance at a future and at life. Another approach would be to invest our money in younger patients regardless of condition.

It is not clear which approach would let us invest our money "effectively according to our preferences," because to define our "preferences" we must consider a number of factors such as equality, maximization, clinical practice, the degree of societal consensus on the value of life, moral imperatives, and national culture. From a technical point of view, we could say that society's "welfare equation" must be fully known and determined to make such a decision. We also need to answer questions such as "for whom is this result better?"; "who will be the one to determine the result?"; and "who will be the one to implement it properly, as part of a leadership with its own needs, possibly its own separate goals, and perhaps a short and difficult political term?"

We avoid further details regarding this issue because they might be tiresome to the reader, but we should point out that this particular step is often overlooked. Frequently, matters of economics are treated with a rough, "modern," and summary mathematical approach whose robust assumptions are expressed in vague scientific terms and equations. The purpose of this comment is to introduce to the reader the notion that economics is primarily a *social* science and that mathematics, although an elegant discipline, cannot possibly consider all the social parameters of a decision involving citizens' health care.

In view of this and to define a typical analytical approach for the QALY, nine different assumptions about it are listed:

1. A decision needs to be made regarding the allocation of resources
2. The results of the various interventions can be measured (health condition or improvement, survival time or duration of disease, etc.)
3. The available resources are limited and the adoption of each intervention entails costs for the system
4. An important goal of the decision-makers is to maximize the population's health based on economic and other restrictions

5. Health is defined in accordance with the length and quality of life (how long and how well one lives)
6. Preferences regarding the states of health can be measured
7. Each patient is neutral regarding risk (risk-neutral) and has a utility additive over time. In practice, the first half of this assumption stipulates that the utility offered to the patient by a potential uncertain treatment that includes favorable and unfavorable outcomes (such as "death" and "complete cure") is equal to the utility provided for the "average" of the intervention. In other words, patients should be displeased by an unfavorable development to the same extent that they are pleased by a favorable development, and those two outcomes are balanced. This might not be the case because people are strongly averse to negative developments, and it is rather unlikely that the possibility of a favorable outcome will offset their displeasure before they make their final decision. However, society may be considered to be neutral with regard to risk, which is a reasonable approach. For example, society is interested in statistical measures of the success of a therapeutic intervention and does not care about which individuals specifically will survive and for how long (in contrast, of course, to the patients themselves who wish to minimize any risks of an intervention with regard to their own lives). The second part of the assumption (that of additivity) implies that the patients will add the utility they receive each day, with no satisfaction or dissatisfaction about their outlook for the rest of their lives. For example, if one patient's remaining life span is 1 week (at the equivalent of "perfect" health) and another patient's remaining life span is 10 weeks (also at the equivalent of "perfect" health), then the second patient's utility will be 10-times that of the first patient, according to the assumption. However, this might not be the case because the second patient has a greater absolute survival reservoir and can therefore make arrangements for several life activities and complete activity cycles. This means that this patient will probably enjoy greater utility than 10-times that of the less fortunate patient.
8. Preferences may be added between individuals. This is also an ambitious assumption. According to this assumption, the welfare of a social group can be expressed collectively by adding the individual levels of welfare. We should keep in mind that often, due to pure irrationality, the evolution of a society takes place through conflict between social

groups, between special-interest groups and the rest of the society, and so on, so the unquestioning acceptance of additivity should be evaluated according to the problem to be solved and with consideration to the available data in each case.

9. A QALY is a QALY regardless of who gains or loses it. The last three axioms are obviously simplistic and research is being done to rebut some of their features so that they will more clearly reflect the real world and the decision of the decision-makers. For example, an ambitious approach would be to establish the "weight" of a QALY obtained from cardiology, oncology, or other field and then additively maximize some type of social welfare equation. In any case, such issues are rather delicate and require broad participation of many societal partners to make decisions based on mathematical models.

In concluding this practical summary, there is another "narrower" approach to the QALY in applied research that is more commonly used but that is less interesting from a theoretical viewpoint. This second approach involves the investigation of differences in the quality of life of patients with comparative treatments for *the same condition*. This type of approach is quite reasonable because many new innovative treatments are not different from older ones with regard to survival but are easier to administer and have fewer side effects. These are all important factors and should be taken into account.

ASSESSING THE QUALITY OF AN ECONOMIC ANALYSIS

The methods described here are important to critically appraise as one plans an economic analysis. Lack of attention to these important issues can result in studies that give misleading or inaccurate results. As economic analyses become increasingly common, concern has been raised about the "...validity, methodological quality, and utility of health economic analyses as well as the potential for bias and misuse" (Ofman et al., 2003). This is a serious problem given that many of the decision-makers who rely on these studies are not adequately equipped to critically evaluate the studies, as evidenced by the results of a survey performed by the European Network on Methodology and Application of Economic Evaluation Techniques (Goetghebeur and Rindress, 1999). This has resulted in the construction of a number of instruments designed to critically assess economic studies, although the utility of such tools is limited by their reliance on subjective items and a lack of testing of the tools

	Questions	Points	Yes	No
1.	Was the study objective presented in a clear, specific, and measurable manner?	7		
2.	Were the perspective of the analysis (societal, third-party payer, etc.) and reasons for its selection stated?	4		
3.	Were variable estimates used in the analysis from the best available source (i.e., randomized control trial—best; expert opinion—worst)?	8		
4.	If estimates came from a subgroup analysis, were the groups prespecified at the beginning of the study?	1		
5.	Was uncertainty handled by (1) statistical analysis to address random events and (2) sensitivity analysis to cover a range of assumptions?	9		
6.	Was incremental analysis performed between alternatives for resources and costs?	6		
7.	Was the methodology for data abstraction (including the value of health states and other benefits) staled?	5		
8.	Did the analytic horizon allow time for all relevant and important outcomes? Were benefits and costs that went beyond 1 year discounted (3%–5%) and justification given for the discount rate?	7		
9.	Was the measurement of costs appropriate and the methodology for the estimation of quantities and unit costs clearly described?	8		
10.	Were the primary outcome measure(s) for the economic evaluation clearly stated and did they include the major short-term was justification given for the measures/scales used?	6		
11.	Were the health outcomes measures/scales valid and reliable? If previously tested valid and reliable measures were not available, was justification given for the measures/scales used?	7		
12.	Were the economic model (including structure), study methods and analysis, and the components of the numerator and denominator displayed in a clear, transparent manner?	8		
13.	Were the choice of economic model, main assumptions, and limitations of the study stated and justified?	7		
14.	Did the author(s) explicitly discuss direction and magnitude of potential biases?	6		
15.	Were the conclusions/recommendations of the study justified and based on the study results?	8		
16.	Was there a statement disclosing the source of funding for the study?	3		
	TOTAL POINTS	100		

Figure 4.10 The quality of health economic studies instrument.
Source: Reprinted from Chiou et al. (2003) with permission.

construct validity (Ofman et al., 2003). This led Chiou et al. (2003) to develop and validate a Quality of Health Economic Studies (QHES) instrument that the authors claim will simplify the assessment of quality of health economic evaluations. The QHES uses 16 explicit elements that are answerable using a yes/no approach combined with a weighted scoring system with a total possible score of 100. A full evaluation of the QHES instrument is beyond the scope of the chapter; however, a copy of the tool is provided here (Figure 4.10).

CONCLUSIONS

The reader will note that the elements included in Figure 4.10 reflect the content of this chapter very well. This tool not only is useful in the critical appraisal of published economic studies but also can serve as a guide in the design of economic studies. Careful attention to the methods outlined in this chapter and the elements from the QHES tool will improve the validity and reduce the bias of economic analyses. This approach will yield results that are more relevant to and trusted by the intended audience.

REFERENCES

Barber, J.A., Thompson, A., 1998. Analysis and interpretation of cost data in randomised controlled trials: review of published studies. BMJ 317, 1195−1200.

Black, W.C., 1990. The CE plane: a graphic representation of cost-effectiveness. Med. Decis. Making 10, 212−214.

Briggs, A.H., O'Brien, B.J., 2001. The death of cost-minimization analysis? Health Econ. 10 (2), 179−184.

Canning, D., 2009. Axiomatic foundations of cost effectiveness analysis. *Working Paper Series.* PGDA Working Paper No.51 <http://www.hsph.harvard.edu/pgda/working.htm>.

Cantor, S., 1994. Cost-effectiveness analysis, extended dominance, and ethics. A quantitative assessment. Med. Decis. Making 14, 259−265.

Chiou, C.F., Hay, J.W., Wallace, J.F., Bloom, B.S., Neumann, P.J., Sullivan, S.D., et al., 2003. Development and validation of a grading system for the quality of cost-effectiveness studies. Med. Care 41 (1), 32−44.

Claxton, K., 1999. The irrelevance of inference: a decision-making approach to the stochastic evaluation of health care technologies. J. Health Econ. 18 (3), 341−364.

Collett, D. (Ed.), 2003. Modelling Survival Data in Medical Research, second ed. Chapman & Hall/CRC Press, Boca Raton, FL.

Detsky, A.S., Naglie, I.G., 1990. A clinician's guide to cost-effectiveness analysis. Ann. Intern. Med. 113 (2), 147−154.

Drummond, M., Sculpher, M., Torrance, G., 2005. Methods for the Economic Evaluation of Health Care Programmes. Oxford University Press, Oxford.

Efron, B., Tibshirani, R.J., 1993. An Introduction to the Bootstrap. Chapman & Hall/CRC Press, Boca Raton, FL.

EuroQol Group Association, EuroQol EQ-5D™. <http://www.euroqol.org/fileadmin/user_upload/Documenten/PDF/Products/Sample_UK__English__EQ-5D-3L_Paper_Self_complete_v1.0__ID_23963_.pdf>.

Goetghebeur, M.M., Rindress, D., 1999. Towards a European consensus on conducting and reporting health economic evaluations—a report from the ISPOR inaugural European conference. Value Health 2, 281−287.

Gray, A., Clarke, P., Wolstenholme, J., 2011. Applied Methods of Cost-Effectiveness Analysis in Healthcare. Oxford University Press, Oxford.

Morrison, G.C., 1998. Understanding the disparity between WTP and WTA: endowment effect, substitutability, or imprecise preferences? Econ. Lett. 59, 189.

National Institute for Health and Clinical Excellence. Guide to the Methods of Technology Appraisal, March 2011. <http://www.nice.org.uk/media/B52/A7/TAMethodsGuideUpdatedJune2008.pdf>.

O'Brien, B.J., Gafni, A., 1996. When do the dollars make sense? Toward a conceptual framework for contingent valuation studies in health care. Med. Decis. Making 16, 288−299.

O'Brien, B.J., Gertsen, K., Willan, A.R., Faulkner, L.A., 2002. Is there a kink in consumers' threshold value for cost-effectiveness in health care? Health Econ. 11 (2), 175−180.

Ofman, J.J., Sullivan, S.D., Neumann, P.J., Chiou, C.F., Henning, J.M., Wade, S.W., et al., 2003. Examining the value and quality of health economic analyses: implications of utilizing the QHES. J. Manag. Care Pharm. 9 (1), 53−61.

Palmer, S., Raftery, J., 1999. Opportunity cost. BMJ 318 (7197), 1551−1552.

Pauly, M.V., 1995. Valuing Health Care Benefits in Money Terms. Cambridge University Press, Cambridge.

Ramsey, S., Willke, R., Briggs, A., Brown, R., Buxton, M., Chawla, A., et al., 2005. Good research practices for cost-effectiveness analysis alongside clinical trials: the ISPOR RCT-CEA task force report. Value Health 8 (5), 521–530.

Sullivan, R., Peppercorn, J., Sikora, K., et al., 2011. Delivering affordable cancer care in high-income countries. Lancet Oncol. 12 (10), 933–980.

Torrance, G.W., 1986. Measurement of health state utilities for economic appraisals: a review. J. Health Econ. 5, 1–30.

WHO guide to cost-effectiveness analysis, 2003. <www.who.int/choice/publications/p_2003_generalised_cea.pdf>.

Willan, A., Lin, D.Y., 2001. Incremental net benefit in randomized clinical trials. Stat. Med. 20, 1563–1574.

CHAPTER 5

Advanced Methodological Aspects in the Economic Evaluation

INTRODUCTION

In the previous chapter we have provided an overview of various technical issues of economic evaluation. In this chapter, we describe some methodological issues involved in the economic evaluation of health services. The handling of each of these factors may be extremely important because it can lead to entirely different evaluation results for similar health technologies. The decision-makers must be aware of the assumptions used in each economic model and understand how the assumptions can impact the results of the model. Only then will they be able to use the analysis results as a practical tool for drafting and implementing political decisions.

MODEL TYPES

Generally speaking, there are two types of analysis used in economic evaluation. The first uses patient data (raw data), when the relevant records are available, to calculate all basic economic quantities. In this case, we use techniques borrowed from statistics (which will be explained in the next chapter). A typical case is economic evaluation performed in parallel with a clinical trial. In the second type, we create models simulating the course of the disease with data from the literature. In certain cases, we might use data both from the literature and from the available clinical trial results. For example, in certain cases of chronic diseases, we would be interested in the economic evaluation performed in parallel with a clinical trial (short-term), but we would also be interested in economic evaluation in the longer-term for the same patients; this would necessitate combining data from clinical studies with data from available databases (registries). We also need to determine how our results would be affected if we combined our clinical trial data with cost data from different countries.

In this case, some kind of modeling is necessary. When speaking about "modeling" or "models," we refer to a visual representation (usually presented graphically) that describes the course of a disease or a clinical activity or its treatment, with all possible intermediate options. Models usually need to be simple so that they can be understood, but they should also be sufficiently complex so that they incorporate certain basic features of reality. In practice, the creation of a model is a matter of experience and skill, but it should also result naturally from the type of question we wish to investigate and should be validated by the clinical scientists who work with the analyst.

The most basic form of a model is the decision tree. Such models include "decisions" (represented as squares), "transition possibilities" (represented as circles), "conclusions" (represented as small triangles pointing to the left), and "alternative options" (represented as branches). Such analyses are used if it is not actually important "when" something happens, but we are more interested in the possible outcomes that could occur. This means that such models do not include time as a distinct parameter. A feature of these models is that options are mutually exclusive (e.g., "surgery" or "conservative treatment," etc.), and the sum of the alternative options should equal 100% (e.g., the "surgery" option has a 5% probability of "death," 25% probability of "complications," and 70% probability of "perfect health;" 5% + 25% + 70% = 100%).

The next type of model is the Markov model. This model consists of specific "health states," "transition possibilities" from one state to another, and "cycles" such as the time scale in which patients are periodically evaluated. Markovian models are used when we would like to determine **when** an event takes place and usually have a long-term scope (e.g., in cardiology studies). For example, a model with three health states, each of which was more severe than the next ("patient [stable disease]", "disease progression," "expired") and a yearly cycle, would calculate how many patients each year would fall into one of these three discrete states. It is obvious that, as time goes by, fewer and fewer individuals would be in the first state and more and more would be in the last state. In fact, in some cases such models only "move" in one direction (e.g., a patient with "disease progression" has 0% probability of returning to "patient") and are called "absorbing models." In the typical models, the probability of transition from one state to the next remains constant over time, which is not the case in the clinical sciences; therefore, mathematical modifications have been applied that address variations in the probability of transition between cycles. This approach is reasonable because the passing of time (at least) generally increases the patient's odds of moving more rapidly to a less favorable state of health (e.g., death by old age), regardless of whether there have been any other adverse developments in the particular disease being examined.

THE SIZE OF THE ICER OR THE λ

As discussed in detail in Chapter 4, the ICER is an index that calculates the additional cost required per unit of benefit when comparing a new treatment with the most effective alternative for therapeutic interventions used for the same disease. It is intended primarily to provide information during the decision-making process in the case of more expensive and more effective treatments, which is the most common scenario. Nevertheless, the ICER calculation by itself does not allow conclusions to be drawn about the cost-effectiveness of the various options. As mentioned previously, such conclusions require a quantitative criterion (measured in €/year or $/year), below which an option is considered effective and above which the option is rejected. According to the neoclassical theory of the economics of welfare, a threshold may be defined for ICER to be used as a criterion in deciding which interventions maximize the effectiveness of any health investment at a given time (Weinstein and Zeckhauser, 1973). This type of neoclassical approach can be generalized and used not only for the comparison of interventions for the same disease but also for the comparison of therapeutic interventions aimed at different diseases. Here, the various health interventions are tabulated in a league table based on the mean or median ICER, and then those with the lowest ICER are financed first, then those with higher ICER, and so on, until the entire budget is expended. For example, in a review of the literature involving 242 studies (Greenberg et al., 2010), median ICER values for five different cancer types were determined (Table 5.1).

Table 5.1 Ranking table for various types of cancer

Cancer type	Median ICER (2008 US$)
Colorectal cancer	22,000
Breast cancer	27,000
Lung cancer	32,000
Prostate cancer	34,500
Hematological malignancies	48,000

According to this model, all the patients with the lowest ICER (colorectal cancer patients) would be financed first, then breast cancer patients, and so on. In such a case, certain expensive treatments would be left without reimbursement because they would be significantly more expensive than others and therefore uneconomical for the society, and resources would be insufficient to cover all needs.

A full utilization of the model would require the existence and awareness of the society's welfare equation, a desire by all citizens to maximize this equation, and the establishment of an independent authority that would implement the citizens' wishes through this fully mathematicized equation. It is, however, intuitively obvious that all of these represent a theoretical construct that does not match the reality or the needs of health care decision-makers, nor does it account for political and special interest influences. Furthermore, it is difficult (and possibly pointless) to compare different forms of cancer, let alone completely different disease types that require different patient treatments, that are associated with different ICER values, and also that have different severities, as mentioned in the previous chapter.

Furthermore, the development of the model assumes that the society is using a fixed budget, that a fixed efficiency of scale in health services, which is not affecting the ICER, may be assumed when a program is scaled up or down, that full information is disclosed for all available treatments, and that the individual programs are completely independent.

In many cases, the assumption of a fixed budget may not be true because governments may invest more resources into an intervention program that can appreciably improve the level of health for the population. If the budget is indeed fixed, then the acceptable ICER will change over time as new treatments are developed, which will alter the relative position of the respective diseases in the league table. According to a different version of the model, the ICER could remain fixed provided the budget is variable. In that case, the introduction of new treatments would have no effect on the ICER because the budget would be adjusted accordingly by the equivalent percentage of the additional burden caused by the introduction of the new interventions.

Furthermore, the assumption of fixed economies of scale may be stipulated only if the fixed cost of a program is zero or negligible, so that any changes in its implementation scale will lead to similar health results. The assumption of full disclosure is a simplification because no health system, insurance carrier, or private enterprise can ever be fully informed. In almost every case, it is not certain what benefits will arise and what will be the full economic consequences of the application of a program until those are verified by real data.

Nevertheless, this type of analysis allows economical evaluation of the interventions, which would prevent unnecessary investments in health care if a medical intervention was not cost-effective. Another advantage of this method is the fact that it offers the necessary information to make decisions at a political level and to reveal any trade-off in cost and effectiveness, as well as the economic burden to society for each individual option. Therefore, this analysis can be a useful information tool that can be combined with other information for the purposes of decision-making but not as a sole comprehensive decision-making criterion independent from society. From a limited and intentionally "naïve" economic viewpoint, according to the invariable practice of the exact science of economics (Friedman, 1953), the purpose of these analyses is to reveal relationships that actually exist (in this case, the added cost of the new intervention necessary to obtain an additional year of life), just like "physical" laws are determined in the paradigm of the exact sciences, and not to deal with ethical, moral, or humanistic issues that are not part of the usual professional activities of a health economist; instead, they involve choices expressed collectively by the society through its institutional representatives or through other informal means of social demands in modern democratic states.

Therefore, an alternative approach is recommended. According to this approach, we establish a society's WTP for 1 additional year of life. If the acceptable WTP limit is predetermined, then all interventions with an ICER lower than the limit would be reimbursed and all the others would be considered ineffective. This model obviously raises ethical issues about the way a society treats its patients because the diseases that are being most exhaustively studied by science at any given time are usually associated with more expensive treatments, innovative medicines, and higher ICERs. At the same time, the arguments against the previous model also hold for this model. In practice, it is difficult (if not impossible) to find clear examples of optimizing behavior, because such a solution is inconsistent with Arrow's theorem for a social choice problem (Arrow, 1963). Generally speaking, there cannot be a social utility function that mathematically describes a maximization behavior for the entire population of a modern democracy, where individualism and moral deviations or deficiencies predominate. In many cases, such decisions regarding the distribution of resources for health care are contingent on each country's individual political and historical background and are not determined by economic models.

The estimation of this indicator remains a subject of extensive debate, and even large organizations such as the UK National Institute for Health and Care Excellence (NICE) have yet to announce a clear decision on its "correct" size (Towse, 2009). In various other countries, however, WTP values have been proposed for the "purchase" of 1 year of life to provide a transparent criterion for this difficult undertaking. According to the World Health Organization, the desired value for the indicator is approximately three-times the average per capita income of the country (Eichler et al., 2004). For the United Kingdom, a value between £40,000 and £60,000 is the maximum accepted value in most cases. A value between $50,000 and $100,000 is considered cost-effective, a value less than €20,000 is considered particularly attractive, and values more than €100,000 are considered particularly costly and are rejected (Devlin and Parkin, 2004).

Figure 5.1 presents the possibility of rejection of a treatment by ICER size based on a logarithmic regression model from the UK data. A value of €40,000−€60,000 has a 50% chance of rejection by NICE. Of course, this indicator can be adjusted for special groups such as children or patients with rare diseases, because the social nature of health care and the social imperative for population-wide coverage regardless of economic status or public expenditures will prevail in such cases. A more recent study has determined the value of λ for various countries, as seen in Figure 5.2.

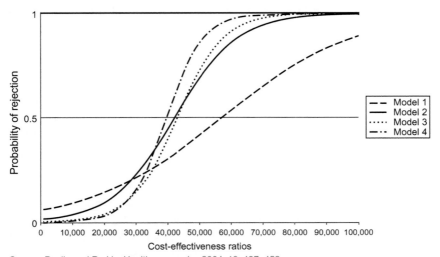

Source: Devlin and Parkin. Health economics 2004; 13: 437–452

Figure 5.1 Probability of acceptance depending on the ICER.

How much does a year of life cost (€)?

Table I Threshold incremental cost-effectiveness ratios
in selected countries

Country	Threshold value in local currency	Threshold value in Euro
Australia	AUS$42,000–$76,000 per QALY	€24,700–€44,700 per life year
Canada	CAN$20,000–$100,000 per QALY	€12,700–€63,300 per QALY
England/Wales	£20,000–£30,000 per QALY	€22,700–€34,100 per QALY
The Netherlands	€20,000–€80,000 per QALY	€20,000–€80,000 per QALY
New Zealand	NZ$ 3000–15,000 per QALY	€1400–€7200 per QALY
United States	US$ 50,000 per QALY	€34,400 per QALY

Notes: Local threshold values were converted into Euro using market exchange rates.
Abbreviation: QALY, quality-adjusted life year.

Figure 5.2 The amount of λ in various countries.

It should be noted, however, that the determination of λ has direct effects on health system budgets and therefore should not be done independently of each economy's available funds.

With regard to λ, we should also keep in mind the following:

- Shadow prices: The λ is supposed to reflect the shadow price or the opportunity cost for health care resources used for the purchase of health programs. Nevertheless, from a wider economic analysis perspective (e.g., the social perspective), it is difficult to calculate the budget precisely, which consequently prevents the exact determination of the λ.
- Budget size: In most cases, we usually simplify by assuming that the size of the budget does not affect the marginal cost of the resources spent for health care. This indicates that the advantages lost in one area by withdrawing resources compared with another area where these resources are invested are the same for each euro invested regardless of sector or level of consumption.
- ICER: Because the λ represents the shadow price of the budget (or the opportunity cost), it should be equal to the ICER for the last health care program selected for funding. However, all health care programs are associated with uncertainty; therefore, a deterministic value for λ can never be defined, only a distribution of values. In fact, considering that new programs are constantly introduced for evaluation and reimbursement, even the distribution of the λ will constantly change.
- The λ as a decision rule: The use of the λ as a decision rule for the reimbursement of health care programs entails two basic assumptions: constant

returns on scale and full divisibility. Constant returns mean that the λ is unrelated to the size of the implemented program. In its simplified form, full divisibility means that we can purchase any health care program on a scale as small as needed; this is a simplified form and full-divisibility formulation. Unfortunately, the conditions described here for the correct calculation of the λ are not yet met by any health system in the world.

ESTIMATES OF EFFECTIVENESS

For economic evaluation, one needs to accurately and reliably estimate the effectiveness of each median intervention. In many cases one can refer to various sources to estimate the effectiveness of a treatment, and each of these sources may have different advantages and disadvantages. The ranking of such sources reflects, in part, the reliability of the data collected and, therefore, the quality of the analysis results.

The best known source is the clinical trial, which belongs to a wider class of studies called "controlled experiments." Clinical trials are scientific experiments involving people who suffer from a disease that assess the difference in response between a new treatment and an alternative, for example, different therapy regimens are selected and compared with each other (http://www. who.int). Clinical trials are conducted in four phases with each one assessing different aspects of the technology undergoing investigation, such as safety, maximum tolerated dose of a drug, effectiveness, and comparison of the drug with other alternative treatments in a large population sample. They are considered the most reliable source of data and are currently an obligatory crucible through which each treatment must pass to be used in the general population. They are characterized by high internal reliability thanks to the very strict patient inclusion criteria and the standardized documentation of the results, they may apply double-blind randomization, and they have high validity thanks to support by modern statistical techniques that minimize sampling errors and bias. Their disadvantage is their low external reliability because general population patients have lower compliance to treatment, may suffer from other concomitant diseases, and may present different characteristics than the study reference population.

The second source of data is meta-analysis. Meta-analysis is a statistical technique aimed at summarizing results obtained from clinical trials (Petiti, 1994). It does not constitute original research but draws its information from the clinical trials included in the analysis. The most important advantage of meta-analysis is that it can draw conclusions at greater statistical power than the individual studies, because it includes greater numbers of patients and can reveal statistically significant relationships. Its disadvantages are that data from studies with different designs must be treated similarly despite differences in study design, and that it incorporates the errors of the individual clinical studies. Also, it does not address the shadow reference problem, and many meta-analyses do not actually fulfill the criteria for a high-quality analysis and degenerate into a simple, misguided statistical technique. Specific methodologies have been designed to minimize these types of problems, but they will not be described here because they are not within the scope of this chapter.

Databases (also known as data warehouses) are repositories of data and records accumulated in the daily operations of large organizations. Depending on the organization, they may include data encompassing much of the care provided to patients or insured persons. They are characterized by high external validity because they include large numbers of subjects, but with the disadvantage that they do not provide detailed information regarding clinical effectiveness and may include inaccurate records. A significant limitation of these organizational databases is the type of data captured. Much of the detailed patient information is captured in text blurbs that require manual review or sophisticated natural language processing that is subject to error and bias. Even the structured data (such as diagnosis and medication codes) are generated for reasons more aligned with reimbursement than accurately reflecting the patient's disease process. Coding systems also have systematic deficiencies that impact the ability to extract needed data. This is particularly true in the case of rare disease. Finally, different organizations may use different structured coding systems that have imperfect matching that interferes with data aggregation (e.g., Canada and most of the rest of the developing world uses ICD-10 but the United States is still using ICD-9). In countries without a national health service, such as the United States, the availability of insurance or claims data is held apart from clinical data and regulations inhibit the ability to aggregate these data, leading to data voids in epidemiologic studies (Prokosch and Ganslandt, 2009; Choi et al., 2013; Meineke et al., 2014).

Medical records are records kept by treating physicians for therapeutic or scientific purposes, and they may be electronic or physical. In countries with sophisticated electronic systems, patient records from different centers may be linked or, more commonly, kept separately at each institution. Their greatest advantage is that they record actual patient data instead of assumptions regarding the administration of drugs or treatments. Their disadvantage is that, in most cases, they do not include detailed information, and the information they do contain may not be in the same format for all patients. In medicine, drugs may be used "off label" (Casali, 2007), a fact that complicates the efforts to investigate the effectiveness of individual therapeutic regimens. Such use may be justified when there is evidence that off-label administration may be effective, even if this has not been proven in long-term, expansive clinical trials. In all of these cases, the use of medical records reduces the degree of uncertainty and provides reliable information.

Clinical databases (also called disease-specific databases or registries) focus on patients with specific conditions and are characterized by large follow-up periods, but they have low external validity and one-dimensional information.

Finally, expert panels (such as the Delphi method) are a qualitative method for determining effectiveness based on the opinions of expert physicians from each field (Cialkowska et al., 2008). Their greatest advantage is that they are very useful when data are insufficient or not available, or when data collection is too expensive. This is important because clinicians must make decisions regarding treatment of their patients even when evidence is absent or insufficient. However, they are characterized by high bias and it is difficult to decide who is an "expert" in any given case. The professional guideline development process in the United States relies heavily on the panel model. A 2011 publication from the Institute of Medicine of the US National Academies entitled "Clinical Practice Guidelines We Can Trust" (Greenfield and Steinberg, 2011) outlines a series of recommendations to ensure that practice guidelines are as reliable as possible. It should be noted that the value of information related to assessing effectiveness has a small life span and it is necessary to evaluate treatments regularly, because the emergence of new information can change current medical assessments of regimens and treatments.

In any case, the new tendency when drafting economic studies is to synthesize all available data into a unit model; in that case, we include clinical study data, literature, and databases to reach specific economic evaluation results.

STUDY PERSPECTIVE AND COST DETERMINATION

The concept of the perspective refers to the institution for which any economic consequences from the alternative therapeutic interventions are valued (Muennig, 2008). The study perspective is a key factor when determining the cost categories that will ultimately be involved in the analysis.

The approach selected will not always be the same; it will differ depending on the purpose of the analysis. For example, if the issue involves distributing resources between various sectors of the economy (such as education, health, defense), then the analysis to be used will be the cost-benefit analysis to determine all the consequences of the relevant options. If, instead, the analysis focuses on distributing resources between different sectors of health care (prevention, treatment, etc.) or between different interventions for the treatment of a specific condition, then the consequences to be measured will be more limited. The selection of analysis method is also affected by the person or institution performing it (patient, hospital, insurance carrier, etc.) and by the availability of relevant data. In practice, economic evaluation is used extensively for "narrow-type" analyses (e.g., alternative interventions within a disease), and less so for interventions involving different sectors of the economy. In this case, economic theory assumptions do not fully reflect the individuals' selections and economic evaluation is commonly used as a tool and not as a complete system of values with which to make the relevant decisions.

For example, if the study concerns an insurance carrier, then the analysis is performed from the perspective of the insurance fund and supplier charges are examined; however, if the study focuses on all possible consequences, then the perspective is social (Table 5.2). The social perspective

Table 5.2 Analysis perspective and cost determination

Fields of analysis perspectives	Determination of economic consequences
Society	All types of medical and nonmedical costs Loss of production and productivity. Invisible (hidden) costs
Social security or private insurance	Charges related to reimbursement of suppliers and users
Health care service providers	Various types of variable costs affecting health care expenditures
Patients	All types of costs, focusing on own payments and loss of productivity
Employer/Company	Charges related to insurance costs and loss of productivity

includes all possible consequences with no regard for who pays for them. If the analysis concerns an employer, then it would include the charges for insurance costs and loss of employee productivity. In this case, costs such as transfer to the hospital, feeding costs, and others, which are covered by the employee, will not be included in the analysis, whereas costs related to loss of productivity or temporary personnel would be included. Currently, the social perspective is considered to be the most appropriate for economic analyses because it includes all of the relevant economic flows within a system. However, this should not be interpreted to mean that the tools of economic analysis are not useful for decision-making from other stakeholder perspectives.

When examining the inclusion of costs, we use a process through which the resources expended for the production of medical interventions, the human-hours consumed for providing health care, and the loss to the productive process due to the patient's inability to work and contribute to the national product are calculated.

Specifically, the concept of cost includes the following (Phillips, 2005):
- Direct costs: The actual cost consumed for the intervention
  - *Direct health care costs*: The cost caused by health care suppliers (the total expenditures for monitoring, treatment, diagnostic tests, medication, etc., which result from the treatment)
  - *Direct nonhealth care costs*: Expenditures arising for the patient as a result of the disease as well as the treatment-seeking process (home help costs, travel expenses, special diet expenses, etc.)
- Indirect cost: Financial losses that are a result of the disease and do not include the costs for providing treatment
  - Indirect cost essentially refers to the loss of productivity because of the disease, either because of work absenteeism or because of reduced productivity (presenteeism)
  - It reflects the value of the goods that the patient could have produced if he/she had not become sick
  - It usually includes lost productivity, free time, time expended by relatives providing assistance, and other time lost
  - Finally, this type of cost includes lost productivity due to premature death
- Invisible (intangible) cost: A term that describes difficult-to-measure consequences of the disease and its treatment
  - It is due to the pain, discomfort, reduced quality of life, or other social or moral consequences of the disease (or its treatment)

It should be noted that the concept of intervention "cost" in economic evaluation refers to the "total resources" expended to treat the disease and is not limited to the cost of a specific technology (e.g., the price of the drug) being valued against an alternative. This cost may vary significantly depending on the institution's perspective, and it can increase considerably as we move to broader analyses of the consequences of the disease. Treatments in oncology and cardiology follow a specific pattern of administration, are given in regimens together with many other drugs, and are associated with toxicity and side effects with various probabilities of occurring and very high management costs. In this case, a simple comparison of the price of two drugs is usually misleading because it does not take into account the effect their administration has on the overall burden to the system through utilization of all the relevant resources, such as hospitalization days and medication given to treat toxicities. A theoretical framework for cost calculation is given in Table 5.3.

Cost measurement in economic evaluation is a process associated with disputes concerning which cost variables should be included and which should not. Any variability in the cost categories included will result in incomparability between different economic evaluation studies. Discussions about costs have led to the development of various guidelines aimed at achieving consistency.

THE DISCOUNT RATE

Discounting is a quantitative technique for adapting the value of future flows to current values. It may appear to be a simple process, but it is actually a major issue in cost-effectiveness analysis, with significant consequences on the valuation of alternative health interventions (Katz and Welch, 1993; Hillman and Kim, 1995; Smith and Gravelle, 2001; Cohen, 2003). Currently, no consensus exists among researchers regarding the appropriate discount rate. As a rule, the selections made by economic institutions at different time points with regard to investing or consuming resources for health care or other sectors are very important, both at the individual and at the national level. Since its beginnings, economics has dealt with this important issue, and a simple and understandable model was developed to study it (Shane et al., 2002). We will not go into an elaborate presentation of the axiomatic foundation of such a model, but we should mention that, thanks to its simplicity, it was quickly developed for use in comparing economic choices made at different time points. In brief, the model could reduce all

Table 5.3 Estimation of the cost of a disease

Direct health care costs	Direct non-health care costs	Indirect costs	Invisible costs
Therapeutic interventions	Patient travel to and from hospital, clinic or doctor's practice	Loss or reduction of income due to temporary, partial or complete disability	Compromised quality of life
Laboratory examinations	Family members' travel and stay expenses	Loss of income and productivity for family members in order to care for the patient at home	Social and professional problems
Medical commodities	Home help and nursing care	Loss or reduction in employee productivity and employer and societal production	Adverse psychosocial circumstances
Use of diagnostic technologies	Own payments by the patient		Disruption of family cohesion
Time spent by medical staff and other health care professionals Stay and servicing			

psychological variables of a technical analysis to a simple coefficient without trying to explain human nature (Shane et al., 2002).

So, using a technical approach, we conclude that flow discounting is justified for two main reasons: price inflation and because economic institutions maintain definite time preferences with regard to holding funds and, consequently, to the cost of investment programs.

Inflation is the rate of price changes over time. Let P be the general price level and let T be a unit of time (usually one calendar year); inflation I would then be defined as:

$$I = \frac{dP/P}{T}$$

Prices will change every year; therefore, any health expenditures made today will have greater value compared with future expenditures. From a different viewpoint, one could claim that as the national product grows over time, a fixed flow of expenditure will account for a smaller percentage of the total income and therefore reflect a smaller burden, which should be calculated by the discount method. In this case, the discount rate reflects the rate of economic growth.

Furthermore, even in a world without inflation, a person would rather have an amount of money today than the same amount at any future time. In other words, they would rather enjoy benefits today than in the future. This approach does not reflect any specific economic theories, but rather it reveals the effect of psychological and social factors outside conventional limits of rational thought in individuals or patients. A person will consider future cash flows uncertain; therefore, short-term cash flows are more significant than long-term flows.

To define in practice the preferred discount rate, one possible approach would take into account the interest rate of a financial product free from market risk, such as long-term Treasury bonds, based on the rates of individual time preferences. A second alternative would be to determine the funds' shadow price, which takes into account the social time preference rate. With this analysis, the social time preference interest rate is determined by the total loss of current private consumption in relation to the long-term flow of benefit that will be obtained by undertaking a public program, specifically the health care intervention program under study. The two methods are associated with advantages and disadvantages. A full theoretical presentation of the respective arguments is beyond the scope of this chapter.

Similar arguments are presented for benefit discounting, which has also caused disputes among economists (Drummond and Torrance, 2005). Those opposed to benefit discounting argue that it is difficult to conceive of a "long-term investment in health" or an exchange of cash flows by individuals or society. They also maintain that flow discount implies that less weight is given to future health care benefits, which means that the current generation is given priority over future generations. The final argument is that people have different understanding and approaches to the monetary cost consumed at different time periods and to the health benefits obtained by health care intervention programs.

On the opposing side, the supporters of the discount technique claim that, after an intervention, the patient will "consume" health care benefits over the long-term; therefore, a discount of future flows appears to be a reasonable approach. Furthermore, nondiscount leads to logical inconsistencies because a health program that provides a small and insignificant perpetual benefit would be considered preferable (if there was no discount), regardless of the level of the initial investment. Additionally, and within reasonable margins, it may be assumed that the people's conduct reveals behaviors that "trade" health over time, such as the avoidance of unhealthy habits (smoking, alcohol consumption, etc.), which deprive the person from a certain utility but transfer an amount of health to the future.

In the relevant literature, the balance leans in favor of discounting and all leading organizations that issue guidelines on economic evaluation use some discount rate for the benefits obtained, along with the cost. The rate is not the same for all countries, thus reflecting a different degree of aversion to future risk or uncertainty. The larger this rate is for costs and benefits, the greater weight is given to short-term consequences, whereas a low or zero rate expresses neutrality toward future risk. In many cases, using a high rate can be dangerous for the public health because many prevention programs have a long-term scope. A high rate might lead to such a program being classified as "ineffective," even if it can protect future generations from a dangerous disease.

To circumvent all problems associated with the rate, a safe solution would be to list the results with no discount and then perform various sensitivity analyses using an interest rate ranging from 0% to 10% for cost and benefits. In practice, adopting such an interest rate is not a purely economic matter, but rather an evaluative judgment made by the society through its institutional representatives. A comparative table of discount rates for various countries within and outside Europe is given in Table 5.4. In most cases, the rate is 3—5% and the sensitivity analysis is between 0% and 10% (International Society for Pharmacoeconomic and Industrial Research; http://www.ispor.com).

Regarding the technical aspect, the present discounted value (PDV) for an investment made at a specific time n is calculated by the following formula:

$$\mathbf{PDV} = \frac{C}{(1+r)^n}$$

where C is the cost, r is the discount rate, and n is the year.

Table 5.4 Comparison of discount rates for cost and benefits in various countries

	Cost	Benefit
Baltic countries	5%	5%
Belgium	3%	1.5%
Canada	5%; SA: 0−3%	5%; SA: 0−3%
Finland	3% and 0%	3% and 0%
France	0%, 3% and 5%	0%, 3% and 5%
Germany	5%, SA (0%, 3%, 10%)	5%, SA (0%, 3%, 10%)
Hungary	5%; SA: 3−6%	5%; SA: 0−6%
Ireland	YES	YES
Italy	3%; SA: 0−8%	3%; SA: 0−8%
New Zealand	3.5%, SA: (0%, 5%, 10%)	3.5%, SA: (0%, 5%, 10%)
Norway	2.5%, SA: 0−8%	2.5−5%, SA: 0−8%
Poland	0% and 5%	0% and 5%
Portugal	5%	5%, 0%
Scotland	6%; SA: 0%—full treasury discount rate	1.5%; SA: 0%—full Treasury discount rate
Spain	6%	6%
Sweden	3%; SA: 0−5%	3%; SA: 0−5%
The Netherlands	4%	1.5%
United Kingdom	3.5%; SA: 0−6%	3.5%; SA: 0−6%
United States	Expert recommendation	Expert recommendation
Switzerland	2.5%, 5%, 10%	2.5%, 5%, 10%
China	Yes and no, with justification	Yes and no, with justification
Austria	5%, SA: 3−10%	5%, SA: 3−10%
Brazil	5%, SA: 0−10%	5%, SA: 0−10%
Cuba	Yes, as recommended by state experts	Yes, with justification
Slovakia	7%	7%
Thailand	3%	3%
Mexico	5%, SA: 3−7%	5%, SA: 3−7%

SA, sensitivity analysis.
Source: Data adapted from http://www.ispor.com, International Society for Pharmacoeconomics and Outcome Research (ISPOR, 2010).

If the cost of the intervention is spread equally over n years, then the formula is generalized as follows:

$$\textbf{PDV} = \frac{C}{(1+r)^1} + \frac{C}{(1+r)^2} + \frac{C}{(1+r)^3} + \cdots + \frac{C}{(1+r)^n}$$

or

$$PDV = C\left(\frac{1}{(1+r)^1} + \frac{1}{(1+r)^2} + \frac{1}{(1+r)^3} + \cdots + \frac{1}{(1+r)^n}\right)$$

or

$$PDV = C\left(\frac{1 - (1+r)^{-n}}{r}\right)$$

An example of the technique's application for a 10-year investment at a rate of 5%, beginning at year 1, is given in Table 5.5.

In this example, an amount of €10,000 invested annually over 10 years is equivalent to €7,722 in present values. The result can be directly calculated using the formula as follows:

$$PDV = C\left(\frac{1 - (1+r)^{-n}}{r}\right) = €1,000\left(\frac{1 - (1+0.05)^{-10}}{0.05}\right) = €7,722$$

For a technical manipulation of inflation and time preferences for a health program, we can either appreciate all future flows using a single, projected percentage of the inflation (provided it is the same for all elements of the cost), or we can appreciate none of the flows and use a lower single interest rate that will express time preferences only. However, if the elements composing the cost have different rates of price change, then either we appreciate them at the appropriate inflation rate and then discount them at a generic rate (inflation + time preference) or we do not

Table 5.5 Calculation of the PDV for a 10-year health investment project[a]

Year	Interest rate	Cost (€)	Discounted cost (€)
1	0.952	1,000	952
2	0.907	1,000	907
3	0.864	1,000	864
4	0.823	1,000	823
5	0.784	1,000	784
6	0.746	1,000	746
7	0.711	1,000	711
8	0.677	1,000	677
9	0.645	1,000	645
10	0.614	1,000	614
Total	**7.722/10 years = 0.722**	**10,000**	**7,722**

[a]($r = 5\%$).

appreciate and we instead use a weighted total discount rate that expresses only time preference. In practice, for reasons of simplicity, prices are taken as fixed (without inflation) and only one rate, expressing time preference between alternative treatments, is used. In many cases, this assumption is methodologically consistent because many interventions have brief time scales and their price changes are negligible. Alternatively, we might say that the discount rate reflects the combined effect of inflation, growth of the economy, and social time preference.

SENSITIVITY ANALYSIS

The term "sensitivity" essentially refers to the way in which our results change when we change our model's assumptions. If sensitivity is high, then the results vary greatly when we change certain assumptions; these assumptions must be very robustly established for our model to have any validity.

To put it concisely and intuitively, we could say that economic evaluation is beset by many kinds of uncertainties. First, there is uncertainty about the structure of the model we have created (structural uncertainty). For example, is our approach to the issue the correct one? Does our analysis reflect reality? Is our model of clinical practice incorrect? To answer such questions, we need to critically evaluate our model and perform various sensitivity analyses, changing its structure to judge its value.

Another source of uncertainty is heterogeneity between the various subgroups to which the model refers (variability due to heterogeneity). If we want to compare the cost and effectiveness of a new cardiology drug with the standard treatment for the management of acute myocardial infarction, then perhaps we should perform separate analyses for men and women, young and elderly patients, patients with previous infarctions, and so on, because these groups will give different results. Because heterogeneity is important for economic models, it should be included in economic evaluation. Such subanalyses (men vs. women, etc.) are sensitivity analyses for heterogeneity.

Uncertainty also exists between patients. We cannot be sure how long a patient will actually live after a procedure, and even "similar" patients do not survive for the same length of time, nor does their care cost the same. This type of uncertainty is called first-order uncertainty. This would theoretically decrease if data from studies with large samples or valid meta-analyses were available. Realistically, there will always be some first-order uncertainty because nature is inherently stochastic.

The uncertainty that is associated with the exact value of a statistical parameter (and that is estimated by the standard deviation) is called second-order uncertainty. This type of uncertainty is of significant interest. For example, if we estimate that the cost of the treatment of a chemotherapy patient is €10,000 ± 2,000 (mean ± standard deviation), then we have entered second-order uncertainty.

Sensitivity analysis in this case is a technique that estimates the effect that different values of an independent variable have on the end results (Jain et al., 2011). Sensitivity analysis is very important when examining the robustness and validity of our conclusions based on the significance of the initial parameters (Meltzer, 2001; Yoder, 2008). It needs to be performed mostly for three purposes:

- First, when the evaluator wants to determine the range of values in which the proposition of the economic model is valid
- Second, to increase the model's reliability when the input data are elastic (e.g., when estimates are used)
- Third, to make the model evaluation convenient to the end user.

The sensitivity analysis methodology consists of three steps. First, the uncertainty parameters are determined. Second, the range of variation is determined. Third, the results are calculated based on the most likely prediction as well as the "direction" of the results. This means that we examine whether one of the interventions is superior to the other according to a certain statistical probability—usually the 95% or 99% significance level.

The most common forms of sensitivity analysis are:

- Single sensitivity analysis: Single analysis explores ICER variations when a single variable of the model—a different one each time—is altered. The variation values are usually within the variation range of the confidence interval, or alternatively they can include all the values found in the literature.
- Multiple sensitivity analysis: A multiple analysis is performed to assess simultaneous changes in two or more variables, such as effectiveness and cost. Similarly, the variation values are obtained from the confidence intervals or determined from the literature.
- Probabilistic sensitivity analysis: Probabilistic sensitivity analysis (PSA) deals with the significant problem of *statistical* estimation of quantities, as in the example of the chemotherapy patient we mentioned previously, and should always be included in any reliable economic analysis. For example, when the variables examined are strongly correlated, are uncertain, or follow distributions, then single sensitivity analysis is not appropriate. The same is true for data originating from different sources.

Based on the reasoning used so far, the ICER was calculated in a deterministic way—with only one point estimate—with no uncertainty when drawing conclusions. In practice, the ICER has a probabilistic nature because the types of costs and the benefit from each intervention follow theoretical or empirical distributions (Briggs and Fenn, 1998). Furthermore, the introduction of a new intervention into the health system entails assuming certain risks, in which case this type of analysis is indicated for handling the uncertainty associated with those options. PSA takes into account the mean value, the standard deviation, and the distribution of each variable, creating thousands of results under computer simulation by selecting random cases based on the defined assumptions. It then summarizes and displays the results obtained, despite its limitations, through the use of the acceptability curve (O'Hagan et al., 2000; Fenwick et al., 2004; Barton et al., 2008).

We present a conceptual example of the necessity and understanding of the method. We assume that the reader is familiar with the concept of distribution. In this simple example, we assume that the cost consists of only one factor (e.g., only the drug). If the cost comprised many factors, then we would need to generalize the method for each one separately and obtain the total sum of the individual distributions.

Let us assume that we are comparing two second-line treatments for lung cancer. Assume that the survival for these two patient groups (and the respective costs) is estimated as follows:

E_T = 10 months
E_S = 9 months
C_T = €10,000
C_S = €9,000

Simple calculations give the ICER based on the mean value, as shown in Figure 5.3.

According to this, the ICER is €1,000 per month or €12,000 per year. If society's willingness to pay (λ) is even slightly higher, such as, €12,001 or €12,002, then the new treatment is considered 100% cost-effective, as can be seen from the diagram (Figure 5.4). The opposite is true if the λ is marginally less than €12,000: there is a 100% chance that the treatment is not cost-effective. It is obvious that such an approach is quite "strict" and does not leave any room for error.

In practice, we will not restrict ourselves to an estimate of the ICER, but we will perform many experiments to calculate more than one ICER based on the distributions available for our variables. We should note that

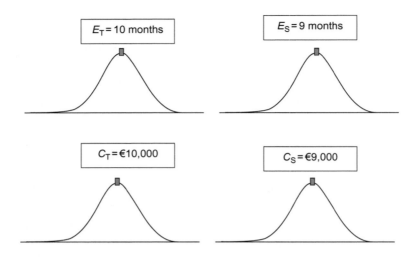

ICER = (€10,000–9,000) / (10–9) = €1,000 per month

Figure 5.3 An example of ICER calculation.

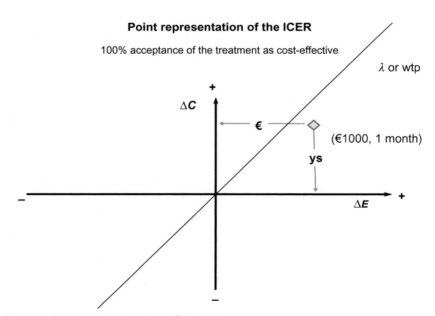

Figure 5.4 Point representation of the ICER.

the selection of the distribution is important, because its characteristics will affect the realization of specific values on cost and effectiveness. Two experiments (out of the thousands we would actually run in the case of a real analysis) are presented in Figure 5.5.

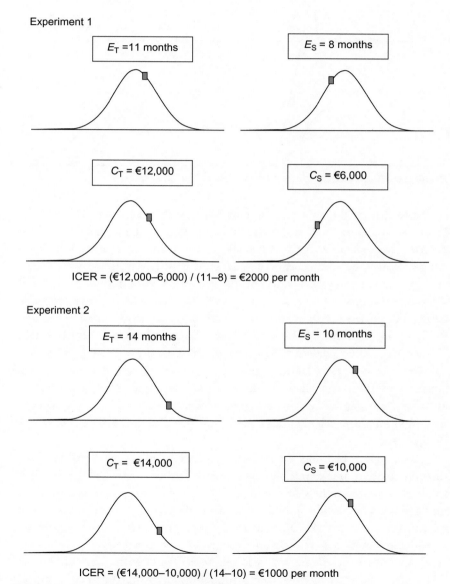

Experiment 1

E_T =11 months E_S = 8 months

C_T = €12,000 C_S = €6,000

ICER = (€12,000–6,000) / (11–8) = €2000 per month

Experiment 2

E_T = 14 months E_S = 10 months

C_T = €14,000 C_S = €10,000

ICER = (€14,000–10,000) / (14–10) = €1000 per month

Figure 5.5 A probabilistic approach to the ICER.

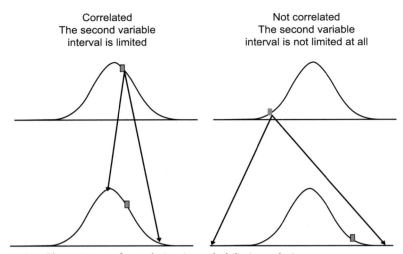

Figure 5.6 The concept of correlation in probabilistic analysis.

Note that in the two experiments we ran, the ΔC, the ΔE, and the ICER are different each time, which means that we have moved on from deterministic results to stochastic analysis (i.e., analysis that includes uncertainty). We should, however, note that in this simple example we did not take into account any correlation in the data, but we arbitrarily assumed that there is no correlation. For that reason, we allow variables to move independently along their entire distribution. In reality, correlation often appears in the data; for example, if the cost is high, then survival may also be high. This concept of correlation is represented schematically in Figure 5.6.

Regardless of correlation (positive or negative), by this process we create many ICERs (as many as 5,000 or more) and plot them in a diagram as shown in Figure 5.7, where each point represents one experiment and, therefore, one calculation of the ICER (the diagram is based on hypothetical results).

This diagram prompts the following question: How can I handle all of this information and represent it concisely? The last step of this analysis is to assume various values of λ within a reasonable range and use these to find the percentage of points that are cost-effective. In this way, instead of relying on an unrealistic approach in which the intervention would be considered 100% cost-effective at exactly €1 above the ICER we calculated and not cost-effective at €1 below it, again with a probability of 100%, the analysis will now take a probabilistic nature. Let A, B, and C symbolize various values of λ, which we assume (in applied analysis the

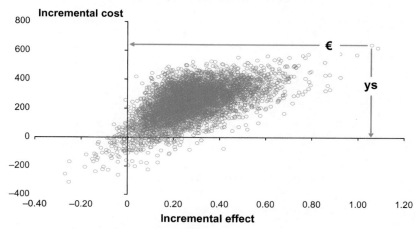

Figure 5.7 Dot plot of ICER representation.

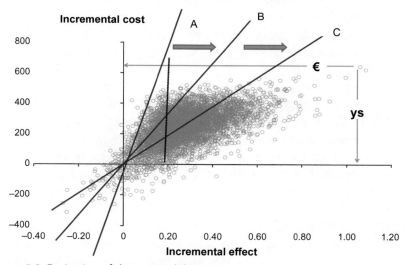

Figure 5.8 Derivation of the acceptability curve.

range is usually €0 to €50,000 or €100,000). The diagram (Figure 5.8) answers the following questions: If the λ is equal to A, then how many points are cost-effective out the total (percentage)? If the λ equals B, then how many points are cost-effective?

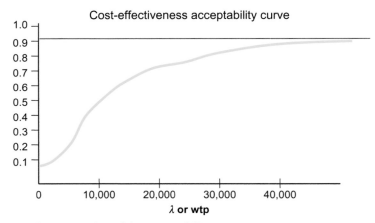

Figure 5.9 Representation of the acceptability curve.

The curve summarizing this information is called the cost-effectiveness acceptability curve (CEAC) and is the output of the probabilistic approach. The scale on the vertical axis is 0–100%, and the horizontal axis represents various values of λ. Based on this curve, we are able to inform budget managers regarding the percentage of probabilistic analysis experiments in which the new treatment is cost-effective compared with the standard as soon as they inform us of their own λ. A hypothetic example of such a curve is shown in Figure 5.9.

It should be noted that plotting a CEAC does not always provide information regarding which treatment is the optimal option and cost-effective compared with an alternative. Under certain conditions (e.g., when comparing more than two options, when the cost and benefit of the treatments follow a specific correlation pattern, when the cost is highly asymmetrical, etc.), a percentage higher than 50% in the CEAC diagram might not indicate a cost-effective option. The curve showing the probability that *the optimal option (the one with the greatest NMB)* is cost-effective for a specific λ is called cost-effectiveness acceptability frontier, and it is what we are most interested in. In this case, the correct way to find the optimal/cost-effective option is to calculate the mean NMB for various values of λ ("find the option with the highest mean NMB for various λ based on the PSA and then find the probability that this is cost-effective"). In this case, there will be areas of λ where one treatment is considered cost-effective and other areas where another treatment is considered cost-effective. Usually, however, when comparing two options in practical research, the CEAC is the correct basis for judging different options.

- Distribution selection in PSA: The success of PSA essentially relies on the correct selection of distributions for each case. Considering that there are standard computer programs with integrated distributions, the selection of a distribution by the investigators may be somewhat arbitrary in some cases. However, if a correct approach is used, consistent with the statistical properties of the data distribution, then the number of possible selections is significantly limited. For example, if we wish to perform a PSA on a probability that is bounded by 0 on the left and by 1 on the right, then we would normally select a suitable statistical distribution like the beta (Claxton et al., 2005). If this probability is derived from the survival analysis coefficients on a log-hazard scale, then selection of the multivariable normal distribution is preferable (Claxton et al., 2005). For right-skewed costs, the gamma or log-scale distributions are commonly used. In cases of data correlation (e.g., differences in cost and benefit), a normal bivariate analysis is usually followed, whereas recent data correlation is also patterned through copula forms (Daggy et al., 2011). We should note that the development of a probabilistic model is indeed a challenge for economists; nevertheless, such an approach is considered more reliable and less arbitrary than simple sensitivity analysis.

THE VALUE OF INFORMATION AND BAYESIAN ANALYSIS

According to the modern approach to economic evaluation, an analysis would ideally use as much of the available information as possible when constructing a model. In addition to study data (e.g., from a new clinical trial), such an approach will also incorporate data from other sources, including the subjective opinion of experts in a scientific field. According to the Bayesian analysis approach, the study we will conduct is simply "one more step on the road to knowledge." This suggests a synthesis of pre-existing knowledge with new knowledge obtained through our own particular study to obtain new, "all-inclusive" knowledge. Such an analysis would first determine our prior perceptions (in the form of "prior" statistical distributions) and combine them with the new data available to obtain the final results: our end perceptions (through posterior distributions). In the literature and in applied research, Bayesian analysis is a new field of study, but the acceptability curve is interpreted more naturally in such a field of analysis.

Let us assume that, ideally, we have created an economic model comparing the cost and effectiveness of two options using some kind of

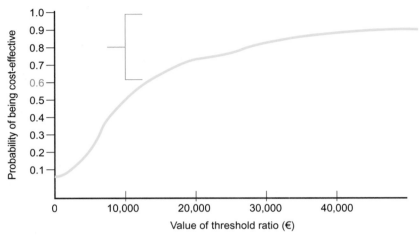

Figure 5.10 The EVPI concept.

Bayesian analysis that incorporates existing societal information with regard to the two therapeutic interventions. Let us also assume that we have performed a PSA and we know that society is willing to purchase an additional year of life at €12,000 ($\lambda = 12{,}000$). As Figure 5.10 shows, the probability that the new option is cost-effective compared with the standard treatment is 60% for $\lambda = €12{,}000$. Although the new treatment should be adopted, we expect that 40% of patients will receive a noncost-effective treatment. In this case, we might be willing—as a society—to allocate some resources for scientific research to convert the 60% estimate to a more "certain" estimate (i.e., closer to 100% or 0%). If the result of further research, based on new data, pushes this percentage closer to 100% (e.g., up to 90% [for $\lambda = 12{,}000$]), then the percentage of patients expected to receive a noncost-effective treatment would be reduced to 10%. If, instead, this probability tended to 0% (e.g., 5%), then the standard treatment should be adopted with probability 95% and the percentage of patients expected to receive a noncost-effective treatment would be reduced to 5%. This means that, in those areas where the probability is close to 50% for a specific λ, it is particularly worthwhile to expend resources to reduce this uncertainty. The value of the resources we are willing to expend to obtain the additional information can be determined objectively and calculated using an analysis called *expected value of perfect information* (EVPI).

We should not confuse our willingness to reduce this uncertainty with the λ. The λ does not add anything to scientific knowledge. It simply indicates how much we are willing to give to "purchase" 1 year of life, and it is taken as fixed here.

At a technical level, what we do is "run" a PSA, convert the difference in benefit between treatments into € based on the λ, and then compare the benefit and cost from each experiment of the 1,000 or 5,000 experiments we usually run in a Monte Carlo simulation (see also previous chapter). If the benefit is negative, then we estimate the total negative benefit and divide it by the number of experiments. One example of an EVPI calculation is given in Table 5.6. Let us say that we are comparing treatments TH1 and TH2 and we know that λ is €10,000/QALY. The PSA included only 20 experiments to simplify the calculations, but the analysis would be the same even if we did 1,000 repetitions. So, assume we ran the 20 repetitions and we obtained the first four columns of the table. The next step is to determine the ΔE (the difference in effectiveness) for each experiment and multiply it by 10,000. Next, we subtract the ΔC (difference in cost) from the ΔE for each experiment and obtain the net monetary benefit (NMB). Note that, out of the 20 experiments, 7 (35%) show that option B, and not option A, is the cost-effective option. We know this now that we have full knowledge of the experiments. How much money would we be willing to pay to avoid this "failure" in 35%? If we add all those who had negative benefit and divide by 20, then we calculate the value of full information. In this example, this value is €639 per patient. By calculating the disease prevalence, the time needed to develop the technology, the discount rates, and other factors, we can determine the total resources we would be willing to invest in research.

We should note that, although the concept of the EVPI is extremely interesting, we unfortunately have to go one step further. Based on the analysis, we found that we would invest up to €639 per patient, BUT the analysis does not say which parameters we should measure to reduce their uncertainty. In this case, we should determine which variables we should give priority to (e.g., we should conduct research to reduce the uncertainty with regard to days of hospitalization for the condition, survival, rate of side effects, etc.). Such an analysis is called *partial EVPI* and is now the leading edge in modern economic evaluation. Mathematic models used in this area are usually complex, and attempts are being made to develop algorithms that will provide reliable results in a reasonable time frame (such an analysis, in its full form, may need several days, even with a fast computer).

Table 5.6 EVPI calculation

	Cost (€) TH1	Cost (€) TH2	Effect TH1	Effect TH12	INB (for λ = 10,000/year)	In this particular experiment, which treatment should I choose based on the perfect knowledge I have now?
Experiment 1	31,898	20,145	4.19	3.1	−802	TH2
Experiment 2	29,570	18,742	3.95	3.19	−3,274	TH2
Experiment 3	30,582	20,028	4.04	2.86	1,294	TH1
Experiment 4	30,915	19,509	4.1	3.04	−793	TH2
Experiment 5	29,416	20,397	3.92	2.98	360	TH1
Experiment 6	28,669	18,784	4.11	2.96	1,602	TH1
Experiment 7	28,827	20,497	3.82	2.92	704	TH1
Experiment 8	28,634	20,157	4.07	3.02	1,994	TH1
Experiment 9	29,575	19,543	3.9	3.06	−1,624	TH2
Experiment 10	29,709	20,756	3.92	3.15	−1,262	TH2
Experiment 11	29,590	19,013	4.51	2.4	10,518	TH1
Experiment 12	29,410	19,621	4.27	2.49	8,091	TH1
Experiment 13	31,507	20,591	4.46	2.88	4,925	TH1
Experiment 14	28,548	19,690	4.09	2.69	5,106	TH1
Experiment 15	28,131	19,771	3.79	2.89	569	TH1
Experiment 16	30,443	21,095	3.94	3.2	−1,968	TH2
Experiment 17	31,165	21,164	3.8	3.1	−3,059	TH2
Experiment 18	30,380	20,126	4.24	2.66	5,510	TH1
Experiment 19	31,124	21,198	4.04	2.86	1,861	TH1
Experiment 20	28,985	20,151	3.73	2.74	1,035	TH1
Average	**29,854**	**20,049**	**4.04**	**2.91**	**€639**	
				EVPI =		

CONCLUSIONS

This chapter, like the previous one, has presented the theory and methodology behind high-quality economic analysis in health care. In the subsequent chapters, we illustrate the application of economic analysis to genomic medicine interventions that are emerging into clinical practice, using some key examples from the published literature. Furthermore, we describe the special provisions required to accurately model the evaluation of the analysis costs and resource use, and the analysis outcomes and effectiveness in genomic medicine applications.

REFERENCES

Arrow, K.J., 1963. Social Choice and Individual Values, second ed. Wiley, New York, NY.

Barton, G.R., Briggs, A.H., Fenwick, E.A., 2008. Optimal cost-effectiveness decisions: the role of the cost-effectiveness acceptability curve (CEAC), the cost-effectiveness acceptability frontier (CEAF), and the expected value of perfection information (EVPI). Value Health 11 (5), 886–897.

Briggs, A.H., Fenn, P., 1998. Confidence intervals or surfaces? uncertainty on the cost-effectiveness plane. Health Econ. 7, 723–740.

Casali, P.G., 2007. The off-label use of drugs in oncology: a position paper by the European Society for Medical Oncology (ESMO). Ann. Oncol. 18 (12), 1923–1925.

Choi, I.Y., Kim, T.M., Kim, M.S., Mun, S.K., Chung, Y.J., 2013. Perspectives on clinical informatics: integrating large-scale clinical, genomic, and health information for clinical care. Genomics Inform. 11 (4), 186–190.

Cialkowska, M., Adamowski, T., Piotrowski, P., Kiejna, A., 2008. What is the Delphi method? Strengths and shortcomings. Psychiatr. Pol. 42 (1), 5–15.

Claxton, K., Sculpher, M., McCabe, C., et al., 2005. Probabilistic sensitivity analysis for NICE technology assessment: not an optional extra. Health Econ. 14 (4), 339–347.

Cohen, B.J., 2003. Discounting in cost–utility analysis of healthcare interventions: reassessing current practice. Pharmacoeconomics 21 (2), 75–87.

Daggy, J.K., Thomas III, J., Craig, B.A., 2011. Modeling correlated healthcare costs. Expert Rev. Pharmacoecon. Outcomes Res. 11 (1), 101–111.

Devlin, N., Parkin, D., 2004. Does NICE have a cost-effectiveness threshold and what other factors influence its decisions? a binary choice analysis. Health Econ. 13 (5), 437–452.

Drummond, M.F., Torrance, G.W., 2005. Methods for the Economic Evaluation of Health Care Programmes. Oxford University Press, Oxford.

Eichler, H.G., Kong, S.X., Gerth, W.C., Mavros, P., Jonsson, B., 2004. Use of cost-effectiveness analysis in health-care resource allocation decision-making: how are cost-effectiveness thresholds expected to emerge? Value Health 7 (5), 518–528.

Fenwick, E., O'Brien, B.J., Briggs, A.H., 2004. Cost-effectiveness acceptability curves—facts, fallacies and frequently asked questions. Health Econ. 13, 405–415.

Friedman, M., 1953. Essays in Positive Economics. University of Chicago Press, Chicago, IL.

Greenberg, D., Earle, C., Fang, C.H., Eldar-Lissai, A., Neumann, P.J., 2010. When is cancer care cost-effective? a systematic overview of cost-utility analyses in oncology. J. Natl. Cancer Inst. 102 (2), 82–88.

Greenfield, S., Steinberg, E., 2011. Clinical Practice Guidelines We Can Trust. Institute of Medicine, National Academies Press, Washington, DC.

Hillman, A.L., Kim, M.S., 1995. Economic decision making in healthcare. A standard approach to discounting health outcomes. Pharmacoeconomics 7 (3), 198–205.

Jain, R., Grabner, M., Onukwugha, E., 2011. Sensitivity analysis in cost-effectiveness studies: from guidelines to practice. Pharmacoeconomics 29 (4), 297–314.

Katz, D.A., Welch, H.G., 1993. Discounting in cost-effectiveness analysis of healthcare programmes. Pharmacoeconomics 3 (4), 276–285.

Meineke, F.A., Stäubert, S., Löbe, M., Winter, A.A., 2014. Comprehensive clinical research database based on CDISC ODM and i2b2. Stud. Health Technol. Inform. 205, 1115–1119.

Meltzer, D., 2001. Addressing uncertainty in medical cost-effectiveness analysis implications of expected utility maximization for methods to perform sensitivity analysis and the use of cost-effectiveness analysis to set priorities for medical research. J. Health Econ. 20 (1), 109–129.

Muennig, P., 2008. Cost Effectiveness Analysis in Health Care, second ed. Jossey-Bass, San Francisco, CA.

O'Hagan, A., Stevens, J.W., Montmartin, J., 2000. Inference for the cost-effectiveness acceptability curve and cost-effectiveness ratio. Pharmacoeconomics 17 (4), 339–349.

Petiti, D., 1994. Meta-Analysis, Desicion Analysis, and Cost-Effectiveness Analysis. Oxford University Press, New York, NY.

Phillips, C.G., 2005. Health Economics. An introduction for Health Professionals. Blackwell Publishing Ltd, Massachusetts.

Prokosch, H.U., Ganslandt, T., 2009. Perspectives for medical informatics. Reusing the electronic medical record for clinical research. Methods Inf. Med. 48 (1), 38–44.

Shane, F., Loewenstein, G., O'Donoghue, T., 2002. Time discounting and time preference: a critical review. J. Econ. Lit. 40 (2), 351–401.

Smith, D.H., Gravelle, H., 2001. The practice of discounting in economic evaluations of healthcare interventions. Int. J. Technol. Assess. Health Care 17 (2), 236–243.

Towse, A., 2009. Should NICE's threshold range for cost per QALY be raised? Yes. BMJ 338 (b181).

Weinstein, M., Zeckhauser, R., 1973. Critical ratios and efficient allocation. J. Public Econ. 2, 147–157.

Yoder, J.L., 2008. The importance of cost inputs and sensitivity analyses in a cost-effectiveness analysis. Clin. Drug Investig. 28 (7), 461–462.

CHAPTER 6

Economic Evaluation in the Genomic Era: Some Examples from the Field

INTRODUCTION

The preceding chapters have explored the theory and methods behind high-quality economic analysis in health care. Although some examples from genomic medicine have been used to illustrate theoretical aspects of the methods, this chapter uses examples from the published literature to examine the application of economic analysis to genomic medicine interventions that are emerging into clinical practice. As has been noted previously in the book, these are early days for the field, so the number of published studies is still relatively small. Nonetheless, the studies that have been published represent a diverse approach to the application of economic analysis to genomic medicine instances that serve the purposes of this chapter well.

This chapter is organized to illustrate some of the opportunities and challenges introduced in the prior chapters. The examples emphasize examination of the critical aspects of the analysis that can impact the validity of the analysis. The chapter finishes with an exploration of the opportunity economic analysis presents in emerging countries looking to implement genomic medicine. The reader should bear in mind that this chapter is not an exhaustive review of the few existing studies of economic evaluation in genomic medicine, but rather a collection of the most representative studies to demonstrate the importance of economic evaluation in genomic-guided therapies.

USING PHARMACOGENOMICS TO PREVENT ADVERSE EVENTS

Adverse drug events (ADE) are a major contributor to morbidity, mortality, and costs of care (Classen et al., 1997, 2010). Considerable effort has been expended to identify preventable causes of ADE and to create systems that reduce risks due to poor reliability of delivery systems (such as computerized order entry). It has long been known that patients respond differently to medications as a result of environmental and individual factors, including genomic variation.

One of the most well-known examples involves the drug abacavir. Abacavir is a synthetic carbocyclic nucleoside analog with inhibitory activity against human immunodeficiency virus (HIV-1). In combination with other antiretroviral agents, it is indicated for the treatment of HIV-1 infection. Serious and sometimes fatal hypersensitivity reactions have been associated with abacavir. Studies of patients who experienced an abacavir-associated ADE identified an association between the ADE and a specific genetic variant in the HLA complex, HLA-B*57:01. Patients who carry the HLA-B*57:01 allele are at high risk for experiencing a hypersensitivity reaction to abacavir (Hughes et al., 2008; Mallal et al., 2008; Saag et al., 2008). Approximately 0.5% of patients who are HLA-B*57:01-negative will develop hypersensitivity, whereas more than 70% who are HLA-B*57:01-positive will develop hypersensitivity. The FDA issued an alert in July 2008 about this and information was added to the box warning (FDA, 2014).

This is a very straightforward case for the application of economic analysis. The first was performed in 2004 by Hughes et al. (2004). The patient level data for abacavir ADE were obtained from a large HIV clinic and the analysis included costs, such as cost of genetic testing, cost of treating abacavir hypersensitivity, and the cost and selection of alternative antiretroviral regimens. The investigators used a probabilistic decision analytic model that compared testing with no testing and tested the model using Monte Carlo simulations. They concluded that based on the choice of comparators, the testing strategy ranged from dominant (less expensive and more beneficial compared with no testing) to an incremental cost-effectiveness ratio (ICER) of €22,811. The study was limited by very small numbers of events, resulting in a very wide range of odds ratios (7.9 with 95% confidence intervals 1.5—41.4). The study was done from the health system perspective because it did not include data to allow for a societal perspective. Several subsequent analyses have been performed, all of which have determined testing prior to the use of abacavir as being cost-effective and potentially cost-saving under some assumptions.

A study by Schackman et al. (2008) used a simulated model of HIV disease based on the prospective randomized evaluation of DNA screening in a clinical trial study. The study modeled three different approaches: (i) triple therapy including abacavir; (ii) genetic testing prior to triple therapy with tenofovir substituted for abacavir for patients who carry the HLA−B*57:01 allele; and (iii) triple therapy with tenofovir substituted for abacavir for all patients. Abacavir and tenofovir were assumed to have equal efficacy, and the cost of the tenofovir treatment was $4 more than the abacavir treatment. Outcomes were quality-adjusted life-years (QALY) and lifetime medical costs. The authors concluded that the genetic testing strategy was preferred and resulted in a cost-effectiveness ratio of $36,700/QALY compared with no testing (the tenofovir strategy was found to increase cost with no improvement in outcomes and thus was dominated). The authors stressed that the model was robust *if* the assumption that abacavir and tenofovir had equivalent efficacy and abacavir therapy were less expensive. This is a very important conclusion because it demonstrates that the result of an analysis is sensitive to changing conditions in the health care system and thus may not remain "true" in the face of these changing conditions.

Kauf et al. (2010) studied this from the US health care system perspective in a very well-defined patient cohort. They used two approaches in their evaluation: (i) a short-term model looking at the efficiency of genetic testing compared with no testing prior to initiation of abacavir and (ii) a lifetime discrete event simulator that compared testing prior to abacavir use versus initiation with a tenofovir-containing regimen. The authors carefully described the sources of the data needed for their analyses and identified gaps in data when expert opinion was used. They used both sensitivity and scenario analyses to assess the impact of parameter uncertainty. For the short-term analysis, the outcome was cost per patient. For the long-term analysis, outcomes were presented as QALY. The results of the short-term analysis showed that while the cost per patient increased slightly by $17 for the tested group, the model predicted that testing would result in the avoidance of 537 ADE per 10,000 patients. In the lifetime model, the testing prior to abacavir use dominated the tenofovir-containing regimen (based on improved effectiveness and decreased cost) in the base case and in sensitivity analyses.

The latter two studies score well on the Quality of Health Economic Studies (QHES) (Ofman et al., 2003) instrument. One can see the challenge that the results could pose to a reader trying to answer the question of whether genetic testing prior to abacavir is cost-effective. It requires the reader to understand the scenarios used for the analysis and to determine the relevance of the results to their specific situation. The fact that all of the studies concluded that testing was cost-effective across a variety of scenarios and assumptions provides some reassurance, although in reality given the FDA recommendation, cost-effectiveness takes the back seat to considerations of patient safety. However, the modeling provides some information about the relative efficiencies of various strategies that could be useful to systems looking to implement the FDA recommendations. This is explored later in the chapter.

BALANCING ADVERSE EVENTS AND EFFICACY

The preceding case was unusual in that the genomic marker was associated solely with a risk for an ADE. In contrast, most medications require balancing the benefits against the potential risks associated with the therapy. The nineteenth century English physician Peter Latham recognized this balance with his pithy quote, "Poisons and medicine are oftentimes the same substance given with different intents." There are a number of factors that can alter an individual's risk/benefit ratio, including germline genomic variants. In this section, we examine some examples of pharmacoeconomic analyses that explore the economic implications of the balance between efficacy and harm.

Thiopurine medications (including 6-mercatopurine [6-MP], 6-thioguanine [6-TG], and azathioprine [AZA]) interfere with purine metabolism and are used for a variety of indications, including acute lymphoblastic leukemia (ALL) and inflammatory and autoimmune diseases, and as immunosuppressants in organ transplant recipients. These medications are metabolized by the enzyme thiopurine methyltranferase (TPMT) encoded by the gene *TPMT*. Variants in the *TPMT* gene that alter the composition of the protein can result in altered enzymatic activity (Relling et al., 2011, 2013). Individuals with reduced or absent TPMT activity are at increased risk for clinically significant myelosuppression compared with those with normal enzyme activity (a harm). This has led to guidelines for reducing the dose and enhancing monitoring for signs of myelosuppression in individuals known to have reduced TPMT activity or who have *TPMT* gene variants known to lead to reduced enzymatic activity (Relling et al., 2011, 2013). Myelosuppression is generally considered an adverse event; however, in the case of ALL, it is the goal of the treatment, in so far as myelotoxicity is needed, to eliminate the malignant lymphoblasts and induce remission. The endpoint of the induction therapy in ALL is elimination of at least 99.9% of blasts from the blood and peripheral bone marrow, which necessitates a significant degree of generalized myelosuppression. In this context, myelosuppression can be looked at as both a beneficial and potentially harmful outcome. Stanulla et al. (2005) examined an important outcome of ALL therapy, minimal residual disease (MRD) in a population of pediatric patients treated with standard therapies that included 6-MP who had undergone *TPMT* genotyping. MRD is the strongest predictor of relapse of ALL, so it is a very important intermediate outcome of treatment. The genotype information collected was not used to adjust the dose of 6-MP. Analysis of the study data showed that patients with a genotype that predicted reduced TPMT activity had an MRD rate of 9.1% compared with 22.8% in the group predicted to have normal TPMT activity (a statistically significant result). This study emphasized the importance of considering outcomes that reflect both efficacy and harm. This caveat is illustrated in the following examples.

One of the earliest published cost-effectiveness studies of *TPMT* genotyping prior to use of 6-MP compared with standard of care was performed by the Institute for Prospective Technological Studies (van den Akker-van Marle et al., 2006). The analysis assumed a genotyping cost of €150 and yielded a mean cost per life-year gained of €2100, which increased to €4300 when the usual 3% discount rate was applied. The authors concluded that *TPMT* genotyping was cost-effective compared with standard care and expected the cost-effectiveness to improve as the genotyping cost was expected to decrease. They recommended that *TPMT* genotyping should be seriously considered for implementation. Let us assume we are the decision-maker who has received this information. What more do we want to understand about the analysis before we act on the recommendation?

Most of the book to this point has focused on the importance of the methods of the analysis, and a careful examination of the methods of this chapter is instructive. One of the first statements in the methods section of this chapter is, "...not much information on the parameters for the *TPMT* model was specifically available for children with ALL." This is a problem given that the analysis is specifically about the use of *TPMT* genotyping in children with ALL. The authors used data from other pharmacoeconomic studies of thiopurines (many in adults with inflammatory disease) and expert interviews. The authors explicitly stated the analysis was performed from the societal perspective but only included direct costs (so costs related to time lost from work, e.g., were not included).

Appropriate price indexing was used and a discount rate for future costs was applied. The authors appropriately considered differences in allele frequency based on ethnicity and included realistic ethnic distributions in the model. The authors made clear that the only complication of treatment considered was the most severe adverse event, myelosuppression. The estimates of the frequency were derived primarily from adult studies in which 6–MP was used for inflammatory bowel disease or other inflammatory conditions. Applying adult experience to pediatric patients is a problem. In addition, the indication for treatment can make a huge difference in the balance of risk and benefit, as is illustrated in the following section of this chapter. A key parameter for the model is the cost of an episode of myelosuppression. The authors identified a number of sources for costs associated with inpatient and outpatient treatments, developing an average cost of the treatment from the perspective of three different national health systems and then making a determination of the average cost to be used for purposes of the model. The sources were cited and the calculations used were provided. Another challenging parameter to determine is how many of the adverse events can be attributed to the presence of a *TPMT* variant. Although presence of decreased TPMT activity is known to be associated with increased risk of myelosuppression, the nature of the treatment means that even those with normal activity are at risk for an adverse event. In fact, given that the majority of patients have normal TPMT activity, the absolute number of adverse events in this population is higher than in the population with decreased activity. The authors concluded that, for the purposes of the model, they would attribute 32% of all cases of myelosuppression to a *TPMT* variant. Stated another way, for every ADE occurring in a patient with decreased TPMT activity, two would occur in patients with normal activity. It is important to state again that the estimates of ADE risk were obtained in studies of *adults* for *disorders other than ALL*. Finally, the authors noted that severe ADEs have the potential to result in loss of life. As could be expected in a study that focuses on children, the prevention of a childhood death leads to a large number of life-years gained. This must be tempered by the fact that ALL can also lead to premature death, so realistically estimating the average life expectancy of all treated patients compared with peers without ALL is very important. We return to this point because it is critical in the balance of risk and efficacy. Given the uncertainty in many of the model parameters, the authors used sensitivity analyses to test the impact of the assumptions made.

In discussing the results, the authors note that *TPMT* genotyping can reduce health care costs through the avoidance of myelosuppression episodes compared with no genotyping. However, these costs saved, "...are a minor consideration compared with the potential for saving lives." Reinforcing this point, the authors noted that a series of univariate sensitivity analyses for the model's parameters had a modest effect on the cost-effectiveness estimates, with the notable exception of mortality prevented. Changing this parameter 10-fold (from prevention of death in 1 in 1,000 treated patients to prevention of death in 1 in 10,000 treated patients) changed the estimates of cost-effectiveness from €2100 per life-year gained to €47,600 per life-year gained. The authors also performed a multivariate sensitivity analysis that gave the boundaries of cost-effectiveness ratios (per life-year gained without applying discounting) between a cost-saving strategy and €53,000. The authors note that even the extreme cases are at or only modestly above the cost-effectiveness threshold usually used (US$50,000 per life-year gained).

This is a well-done high-quality economic analysis, so the conclusion that *TPMT* genotyping should be routinely implemented is reasonable. However, a key assumption is made that is not critically considered in the analysis: the impact of a potential reduction in efficacy. The authors assumed that the alterations in management would reduce ADEs with no impact on the treatment response for the primary disease, ALL. As noted previously (Stanulla, 2005), this assumption may not be correct. If we assume that the rate of MRD is approximately twice that for patients with a wild-type *TPMT* genotype compared with a variant genotype, then a much larger number of patients would experience relapse based on their MRD, thus incurring increased medical costs as a result of more intensive therapies, including bone marrow transplantation. More importantly, this could result in more children who succumb to their underlying disease. The authors did not provide the ALL mortality rate used for the model because it was assumed to be the same in both groups. If we use a long-term event-free survival rate of 90% for children approximately 8 years of age (as in the study), that would mean that out of a cohort of 100,000 treated ALL patients we would expect 10,000 to die from their disease (for the purposes of this exercise, we assume that none of these deaths are due to ADE). Using the study estimate of prevention of 1 death due to an ADE in 1,000 treated patients, we expect that the genotyped cohort would experience 9,900 deaths (10,000 deaths—100 deaths prevented by genotyping). However, if we assume that efficacy of treatment is decreased by 10%, then the mortality rate would increase from 10% to 11%. This means that the number of deaths in the genotyped group would now be 10,900 (11,000 deaths—100 deaths prevented by genotyping), 900 more deaths than in the nongenotyped group. The impact on efficacy would have to be reduced to a mere 1% difference for the number of deaths to be equal in the two groups, a number that seems unrealistically low given the potential impact on MRD. Clearly, the truth is "still out there;" however, this should illustrate the importance of considering both efficacy and harm for the purposes of economic analyses.

THE DISEASE MAKES A DIFFERENCE

Staying with the example of thiopurines and *TPMT*, this section examines the impact that a different treatment indication has on the economic analysis. Thiopurines are a mainstay of treatment for inflammatory diseases, with AZA being the preferred member of the class. We look at an analysis performed in the context of Crohn's disease (Dubinsky et al., 2005) and compare that with the ALL example.

This study examines both efficacy and prevention of adverse events. In contrast to ALL, myelosuppression is not an expected outcome of treatment but it represents a pure adverse event with no associated efficacy. However, if the focus is only on prevention of the adverse event, then conservative dosing could decrease treatment efficacy. Therefore, there still exists a balance between efficacy and adverse events, but the tolerance for myelosuppression in inflammatory disease is lower than in ALL, shifting the balance toward prevention of the adverse event. *TPMT* genotyping is used to identify individuals who have an increased susceptibility to myelosuppression, whereas monitoring of white blood cell counts and disease activity scales were used to assess the efficacy of treatment. Three different approaches were compared with standard of care: (i) *TPMT* genotyping alone; (ii) intensive monitoring; and (iii) *TPMT* genotyping with intensive monitoring. Values for the decision model were obtained from standard resources (for costs), a literature review, and expert opinion when data were not available. The perspective was that of a third-party payer. One- and two-way sensitivity analyses were used to test the model. All three of the alternative treatments provided higher efficacy at a lower cost than the standard of care.

The *TPMT* genotyping-only arm provided the lowest cost of the three alternative treatments, a result that persisted in the sensitivity analyses.

The model did not identify a significant difference related to adverse events, which the authors attribute to a low event rate. Interestingly, the efficacy was similar for all three of the alternative therapies, which was unexpected given that the monitoring and *TMPT* genotyping plus monitoring arms allowed response-based dosing of AZA, which would be expected to result in higher efficacy. It was not clear from the article if the assumptions for efficacy were based on published data (ideally from prospective clinical trials) or relied on expert opinion. These two results raise some questions about the assumptions used in the model, an issue that the authors acknowledge as a weakness given the reliance on expert opinion as opposed to clinical trial data. The lack of robust clinical trial evidence represents an ongoing challenge for economic analysis, as evidenced by the fact that multiple analyses over nearly 15 years still have not yielded a definitive answer regarding the cost-effectiveness of *TPMT* genotyping prior to the use of thiopurines for any indication.

PERSPECTIVES ON PERSPECTIVE

Economic analyses must specify from what perspective they are being performed. The perspective used is important if the result is to be of use to the intended stakeholder. To illustrate this, we use the example of tumor-based screening for Lynch syndrome.

Colorectal cancer (CRC) is one of the most common cancers worldwide. Approximately 2–5% of CRC is due to germline mutations in one of several DNA mismatch repair genes, a condition called Lynch syndrome. Individuals who carry such a mutation have a lifetime risk of developing CRC that approaches 80%, and they are at higher risk for several other cancers. Identification of a mutation carrier allows initiation of earlier and more frequent colonoscopy and consideration of prophylactic surgery, which has been shown to reduce the risk of developing certain types of cancer associated with Lynch syndrome. Identification of an index case allows identification of at-risk family members who have not yet developed cancer. The most effective way to identify Lynch syndrome in a patient presenting with CRC is to screen the tumor for changes that are indicative of a mutation (e.g., absence of staining for one of the mismatch repair proteins or DNA microsatellite instability). Tumor-based screening was recommended based on a systematic evidence review and synthesis by the Evaluation of Genomic Applications in Practice and Prevention Working Group (EGAPP, 2009). Soon thereafter, economic analyses (Mvundura et al., 2010) were performed that demonstrated that universal tumor-based screening for Lynch syndrome was cost-effective from a societal perspective if at-risk family members were identified and screened.

Based on this evidence, the chapter author's institution elected to implement tumor-based screening. However, questions arose about the most cost-efficient way to screen—a question that the published analyses did not address. We constructed decision analytic models and used a combination of budget impact and cost-consequences (or cost-comparison) analyses performed from the perspective of our local health care delivery system (Gudgeon et al., 2011). The primary outcomes of these analytic models included the absolute and relative effectiveness and costs of the different screening protocols by determining intermediate endpoints such as total costs of testing, expected number of cases detected, and cost per case detected. Secondary outcomes, to be defined by stakeholders as we proceeded through decision-making were expected to include various budget impact metrics.

Test performance values were taken primarily from the published literature, although additional data from the referral laboratory (cost and test dependence), our institution's electronic data warehouse (local patient characteristics), and gray data made available by a large research project were used as needed. To estimate specific outcomes pertinent to local decision-makers, we began the analyses open to all testing protocols we believed to be reasonable. Given the rapidly changing nature of the testing protocol, the team monitored new developments and, when appropriate, added new testing protocols to the model. This proved to be a critical decision because the most efficient protocol incorporated an emerging test. This emphasizes the importance of not "locking down" models in a field that changes as rapidly as genomics.

The screening approach with the maximum sensitivity is to sequence the four mismatch repairs in all patients with CRC who have a positive screening test of the tumor. Although this will detect the maximum number of patients with Lynch syndrome, the costs are prohibitive (poor efficiency). The preferred screening approach had only a slight decrease in sensitivity compared with sequencing, but the cost per case detected was dramatically different (sequencing, US$13,355; immunohistochemistry with reflex testing, US$10,412).

Local stakeholders were interested in additional metrics to help them determine the expected impact on clinical resources and budget, including the number of CRC patients treated annually within our delivery system and the number for which we expect to have tumor tissue available for screening. Our models allowed us to estimate the percent of total test costs that would fall under bundled hospital payments (approximately 75%) and percent of screened patients who would be eligible for sequencing, thus requiring consent and genetic counseling (5—6%). After implementation of screening, a question arose as to whether screening should be restricted to patients aged 50 years or younger, given that Lynch syndrome usually results in tumors appearing at a younger age. Models were used to assess the sensitivity of screening and cost per case detected using a variety of age cutoffs. Applying cutoffs reduced the cost of the screening program with a concomitant decrease in sensitivity. Presentation of the data allowed leadership to consider both costs and clinical impact, and it was ultimately decided not to impose an age cutoff for screening (Gudgeon et al., 2013). Finally, the health care system collects data from the screening of all of the key parameters of the model, allowing testing of the performance of the screening program, allowing for continuous improvement.

DEFINING PERFORMANCE CHARACTERISTICS

New technologies are frequently introduced into practice with limited evidence regarding utility. This not only leads to questions about whether the technology is ready to be used but also renders cost-effectiveness analysis heavily reliant on assumptions. One way to address this issue is to create a model that tries to answer the question, what is the magnitude of improvement of the new technology compared with current care needed to meet a defined threshold of cost-effectiveness? We recently performed a threshold analysis in the context of chronic hepatitis C.

Chronic hepatitis C is a major public health issue. Failure to achieve sustained viral response (SVR) with treatment can lead to end-stage liver disease, necessitating expensive interventions such as liver transplantation. Viral genotype is a strong predictor of treatment response, with hepatitis C virus (HCV) genotype 1 showing relative resistance to treatment. It has also been shown that variants in the patient's *IL28B* gene can also predict treatment response, with certain variants leading to a less favorable response. Treatment in HCV genotype 1 has been dramatically improved through the addition of protease inhibitors to standard of care, with the rate of SVR in treatment-naïve patients increasing from 40—50% to

70–80%. Multiple cost-effectiveness (CE) studies have been completed for the new regimens. A systematic review found 11 CE studies of triple therapy in treatment-naïve and pretreated patients with genotype 1 overall and using *IL28B* guidance for better treatment response, with all study results (most from the payer/health system perspective) supporting its cost-effectiveness (San Miguel et al., 2014).

Response rate in HCV genotypes 2 and 3 are much higher than for genotype 1, with SVR being achieved in approximately 80% of patients treated with standard therapy. As with HCV genotype 1, the *IL28B* status of the patient can affect treatment response, although to a lesser degree. Despite the high response rate, there is still a substantial minority of patients who do not achieve SVR and are at risk for development of end-stage liver disease. This represents a potential opportunity for the use of protease inhibitors, but can the cost be justified? Could the *IL28B* status be used to identify patients more likely to experience failure of standard therapy? This was recently modeled (Bock et al., 2014).

This analysis used cohort simulations on decision tree models to estimate SVR rates, treatment costs, and outcomes for multiple treatment arms that were incorporated into a Markov model to predict triple therapy treatment costs, short-term and long-term health care costs, quality-adjusted life-expectancies, cost-effectiveness thresholds, and the relative percent changes in observed SVR rates needed to reach the cost-effectiveness thresholds. The results specify ranges of improvement in SVR rates needed for the *IL28B*-guided triple therapy to be cost-effective for each *IL28B* variant compared with the SOC therapy for HCV genotype 2 and genotype 3 patients. Because the estimated ranges of improvement in SVR rates for each *IL28B* variant are relatively small in comparison with current response rates with standard therapy, they support the study's conclusion that *IL28B*-guided triple therapy for HCV genotype 2 and genotype 3 patients is likely to be cost-effective when long-term outcomes are considered.

COST-EFFECTIVENESS ANALYSIS IN GENOMIC MEDICINE AND THE DEVELOPING WORLD

Although cost-effectiveness studies have mostly been done from the perspective of developed countries, an argument can be made that the analyses could be much more important to developing countries where the consequence of expending scarce resources on technology that is not cost-effective has a much higher opportunity cost (Mitropoulos et al., 2011). Unfortunately, there is also a shortage of health economists and modelers in these countries.

In a recent economic evaluation study involving elderly atrial fibrillation patients using warfarin treatment, it was shown that 97% of elderly Croatian patients with atrial fibrillation belonging to the pharmacogenomics-guided group did not have any major complications, compared with 89% in the control group, and, most importantly, the ICER of the pharmacogenomics-guided versus the control groups was calculated to be just €31,225/QALY (Mitropoulou et al., 2015). These data suggest that, contrary to other developed countries, pharmacogenomics-guided warfarin treatment represents a cost-effective therapy option for the management of elderly patients with atrial fibrillation in Croatia, which may very well be the case for the same and other anticoagulation treatment modalities in neighboring countries.

Apart from the differences in current drug prices and resource utilization in different countries, another important parameter to determine the cost-effectiveness of a certain medical intervention in different health care systems is the variable frequencies of the pharmacogenomic biomarkers. As such, one should bear in mind that a pharmacogenomics-guided medical intervention that is not cost-effective in a certain country may be cost-effective in another country, even if no significant cost differences exist between these two countries because of the higher frequency of a pharmacogenomic biomarker in the general population.

For example, the *CYP2D6* 7* allele frequency is significantly higher in the Maltese population compared with the Caucasian average, whereas the same is true for the *CYP2C9* 3* allele frequency in the Serbian population (unpublished data). This suggests that there may be significant implications in the cost-effectiveness of risperidone and warfarin treatments, respectively, in these two countries. Another example is screening for the *HLA-B* 15:02* allele, which increases the risk of developing Stevens–Johnson syndrome (SJS) and toxic epidermal necrolysis (TEN) when treated with the antiepileptic drugs (AEDs) carbamazepine (CBZ) and phenytoin. A recent systematic review and meta-analysis including 16 studies found considerable variation among different racial/ethnic populations in the relationship between *HLA-B* 15:02* and CBZ-induced SJS and TEN, as illustrated by a summary odds ratio of 79.84 (95% CI, 28.45−224.06), with the following strong relationships warranting screening for three racial/ethnic subgroups: (i) Han-Chinese, 115.32 (18.17−732.13); (ii) Thai, 54.43 (16.28−181.96); and (iii) Malaysian, 221.00 (3.85−12,694.65). Among individuals of white or Japanese race/ethnicity, no patients with SJS or TEN were carriers of the $HLA\text{-}B^*15{:}02$ allele (Tangamornsuksan et al., 2013). This variability can impact estimates of cost-effectiveness even in a single country. An example is a cost-effectiveness analysis of three ethnic groups in Singapore with different allele frequencies of $HLA\text{-}B^*15{:}02$. Genotyping for $HLA\text{-}B^*15{:}02$ and providing alternate AEDs to those who test positive were found to be cost-effective for Singaporean Chinese and Malays, but not for Singaporean Indians, based on ICERs of \$37,030/QALY for Chinese patients, \$7,930/QALY for Malays, and \$136,630/QALY for Indians (Dong et al., 2012).

This suggests that pharmacogenomic economic evaluation studies must be replicated in every country to inform policymakers prior to the implementation of a pharmacogenomic-guided medical intervention to evaluate its cost-effectiveness based on characteristics specific to each country (Snyder et al., 2014). This is a daunting expectation, given the rapidly increasing number of pharmacogenomics guidelines of developed country regulatory agencies such as the US Food and Drug Administration and multinational organizations such as the Clinical Pharmacogenetics International Consortium, and given that many developing countries do not have the necessary resources and expertise to perform the analyses. This problem could be minimized by the construction of generic economic models that would allow input of certain key variables such as allele frequency, test, and treatment costs that vary by country. This model could be used by less experienced individuals who have access to the country-specific variables to generate a first approximation of cost-effectiveness, allowing prioritization between different emerging tests. The decrease of genotyping costs for the once-in-a-lifetime determination of an individual's personalized pharmacogenomics profile using next-generation sequencing technologies including whole-genome sequencing (Mizzi et al., 2014) would gradually result in a cost-effective pharmacogenomics-guided treatment for drugs bearing pharmacogenomic testing recommendations on their labels, because the single test could provide information that would inform prescribing choices over the lifetime of the patient.

CONCLUSIONS

Genomic cost-effectiveness analysis has the potential to inform assessments about the value of current and emerging technologies and prioritize value-based decisions about adoption and investment. Efforts to close evidence gaps can strategically target areas of greatest need and potential health and cost impacts. Economic evaluations from the perspective of the relevant stakeholder can provide information and guidance for decision-making and policy-making. Results from these evaluations can include estimated ranges and threshold levels for key outcome variables to achieve desirable real-world results. Economic analyses should adhere to quality standards such as the QHES tool. Most importantly, given the rapidly changing nature of the evidence in genomics, economic models must be developed that are flexible and adaptable. Ultimately, high-quality and robust models must be developed that can be utilized by experienced stakeholders without high-level training in economics to encourage routine use of these models to assist in decision-making.

REFERENCES

Bock, J.A., Fairley, K.J., Smith, R.E., Maeng, D.D., Pitcavage, J.M., Inverso, N.A., et al., 2014. Cost-effectiveness of IL28B genotype-guided protease inhibitor triple therapy versus standard of care treatment in patients with hepatitis C genotypes 2 or 3 infection. Public Health Genomics 17, 306–319.

Classen, D.C., Pestotnik, S.L., Evans, R.S., Lloyd, J.F., Burke, J.P., 1997. Adverse drug events in hospitalized patients. Excess length of stay, extra costs, and attributable mortality. JAMA 22 (277), 301–306.

Classen, D.C., Jaser, L., Budnitz, D.S., 2010. Adverse drug events among hospitalized medicare patients: epidemiology and national estimates from a new approach to surveillance. Joint Comm. J. Qual. Patient Saf. 36 (1), 12–21.

Dong, D., Sung, C., Finkelstein, E.A., 2012. Cost-effectiveness of HLA-B*1502 genotyping in adult patients with newly diagnosed epilepsy in Singapore. Neurology 79, 1259–1267.

Dubinsky, M.C., Reyes, E., Ofman, J., Chiou, C.F., Wade, S., Sandborn, W.J., 2005. A cost-effectiveness analysis of alternative disease management strategies in patients with Crohn's disease treated with azathioprine or 6-mercaptopurine. Am. J. Gastroenterol. 100, 2239–2247.

Evaluation of Genomic Applications in Practice and Prevention (EGAPP) Working Group, 2009. Recommendations from the EGAPP working group: genetic testing strategies in newly diagnosed individuals with colorectal cancer aimed at reducing morbidity and mortality from lynch syndrome in relatives. Genet. Med. 11, 35–41.

FDA Information for Healthcare Professionals: Abacavir (marketed as Ziagen) and Abacavir-Containing Medications. <http://www.fda.gov/Drugs/DrugSafety/PostmarketDrugSafetyInformationforPatientsandProviders/ucm123927.htm> (accessed 31.07.14).

Gudgeon, J.M., Williams, J.L., Burt, R.W., Samowitz, W.S., Snow, G.L., Williams, M.S., 2011. Lynch syndrome screening implementation: business analysis by a healthcare system. Am. J. Manag. Care 17, e288–e300.

Gudgeon, J.M., Belnap, T.W., Williams, J.L., Williams, M.S., 2013. Impact of age cutoffs on a Lynch syndrome screening program. J. Oncol. Pract. 9, 175–179.

Hughes, C.A., Foisy, M.M., Dewhurst, N., Higgins, N., Robinson, L., Kelly, D.V., et al., 2008. Abacavir hypersensitivity reaction: an update. Ann. Pharmacother. 42, 387–396.

Hughes, D.A., Vilar, F.J., Ward, C.C., Alfirevic, A., Park, B.K., Pirmohamed, M., 2004. Cost-effectiveness analysis of HLA B*5701 genotyping in preventing abacavir hypersensitivity. Pharmacogenetics 14, 335–342.

Kauf, T.L., Farkouh, R.A., Earnshaw, S.R., Watson, M.E., Maroudas, P., Chambers, M. G., 2010. Economic efficiency of genetic screening to inform the use of abacavir sulfate in the treatment of HIV. Pharmacoeconomics 28, 1025–1039.

Mallal, S., Phillips, E., Carosi, G., Molina, J.M., Workman, C., Tomazic, J., et al., 2008. PREDICT-1 study team. HLA-B*5701 screening for hypersensitivity to abacavir. N. Engl. J. Med. 358, 568–579.

Mitropoulos, K., Johnson, L., Vozikis, A., Patrinos, G.P., 2011. Relevance of pharmacogenomics for developing countries in Europe. Drug Metabol. Drug Int. 26, 143–146.

Mitropoulou, C., Fragoulakis, V., Bozina, N., Vozikis, A., Supe, S., Bozina, T, et al., Economic evaluation for pharmacogenomic-guided warfarin treatment for elderly croatian patients with atrial fibrillation. In Press 2015.

Mizzi, C., Mitropoulou, C., Mitropoulos, K., Peters, B., Agarwal, M.R., van Schaik, R.H., et al., 2014. Personalized pharmacogenomics profiling using whole genome sequencing. Pharmacogenomics 15, 1223–1234.

Mvundura, M., Grosse, S.D., Hampel, H., Palomaki, G.E., 2010. The costeffectiveness of genetic testing strategies for lynch syndrome among newly diagnosed patients with colorectal cancer. Genet. Med. 12, 93–104.

Ofman, J.J., Sullivan, S.D., Neumann, P.J., Chiou, C.F., Henning, J.M., Wade, S.W., et al., 2003. Examining the value and quality of health economic analyses: implications of utilizing the QHES. J. Manag. Care Pharm. 9, 53–61.

Relling, M.V., Gardner, E.E., Sandborn, W.J., Schmiegelow, K., Pui, C.H., Yee, S.W., et al., 2011. Clinical Pharmacogenetics Implementation Consortium. Clinical pharmacogenetics implementation consortium guidelines for thiopurine methyltransferase genotype and thiopurine dosing. Clin. Pharmacol. Ther. 89, 387–391.

Relling, M.V., Gardner, E.E., Sandborn, W.J., Schmiegelow, K., Pui, C.H., Yee, S.W., et al., 2013. Clinical Pharmacogenetics Implementation Consortium guidelines for thiopurine methyltransferase genotype and thiopurine dosing. Clin. Pharmacol. Ther. Advance online publication 20 February 2013. http://dx.doi.org/10.1038/clpt.2013.4.

Saag, M., Balu, R., Phillips, E., Brachman, P., Martorell, C., Burman, W., et al., 2008. Study of hypersensitivity to abacavir and pharmacogenetic evaluation study team. High sensitivity of human leukocyte antigen-B*5701 as a marker for immunologically confirmed abacavir hypersensitivity in white and black patients. Clin. Infect. Dis. 46, 1111–1118.

San Miguel, R., Gimeno-Ballester, V., Mar, J., 2014. Cost-effectiveness of protease inhibitor based regimens for chronic hepatitis C: a systematic review of published literature. Expert Rev. Pharmacoecon Outcomes Res. 14, 387–402.

Schackman, B.R., Scott, C.A., Walensky, R.P., Losina, E., Freedberg, K.A., Sax, P.E., 2008. The cost-effectiveness of HLA-B*5701 genetic screening to guide initial antiretroviral therapy for HIV. AIDS 22, 2025–2033.

Snyder, S.R., Mitropoulou, C., Patrinos, G.P., Williams, M.S., 2014. Economic evaluation of pharmacogenomics: a value-based approach to pragmatic decision-making in the face of complexity. Public Health Genomics 17, 256–264.

Stanulla, M.(1), Schaeffeler, E., Flohr, T., Cario, G., Schrauder, A., Zimmermann, M., et al., 2005. Thiopurine methyltransferase (TPMT) genotype and early treatment response to mercaptopurine in childhood acute lymphoblastic leukemia. JAMA 293, 1485—1489.

Tangamornsuksan, W., Chaiyakunapruk, N., Somkrua, R., Lohitnavy, M., Tassaneeyakul, W., 2013. Relationship between the HLA-B*1502 allele and carbamazepine-induced Stevens-Johnson syndrome and toxic epidermal necrolysis: a systematic review and meta-analysis. JAMA Dermatol. 149, 1025—1032.

van den Akker-van Marle, M.E., Gurwitz, D., Detmar, S.B., Enzing, C.M., Hopkins, M.M., Gutierrez de Mesa, E., et al., 2006. Cost-effectiveness of pharmacogenomics in clinical practice: a case study of thiopurine methyltransferase genotyping in acute lymphoblastic leukemia in Europe. Pharmacogenomics 7, 783—792.

CHAPTER 7

Special Requirements for Economic Evaluation and Health Technology Assessment in Genomic Medicine

INTRODUCTION

From the previous chapter, it is obvious that genomic interventions could enable improved patient stratification and tailor-made treatment interventions. To date, however, most genetic tests focus on a gene-by-gene approach, rarely detecting multiple genomic variants at high resolution and accuracy. Technological breakthroughs have led to the development and early adoption of genomic tests with the potential to meet these criteria, namely whole-exome and whole-genome sequencing. The latter technologies can simultaneously detect at a genome-wide scale all of an individual's genomic variants and, as such, have huge potential in genomic medicine (Mardis, 2008; Mizzi et al., 2014). Hybrid technologies such as next-generation sequencing panels sequence multiple genes known to be associated with a given condition, such as hereditary cancer (Laduca et al., 2014) or intellectual disability (Flore and Milunsky, 2012). These have the advantages of improved sequencing quality and depth compared with whole-exome and genome sequencing while minimizing issues related to incidental findings (ACMG IF report). As an example for pharmacogenomic testing, PGRNSeq (designed by the Pharmacogenomics Research Network) is a next-generation sequencing platform that assesses sequence variation in 84 proposed pharmacogenes (Gordon et al., 2012).

To date, only a small fraction of drugs have pharmacogenomic information on their labels and are approved by the US Food and Drug Administration (http://www.fda.gov) and the European Medicines Agency (http://www.ema.europa.eu). Also, there is a variable rate of adoption of genomics in various countries worldwide, and this rate is lower in developing countries with limited resources. Such discrepancies arise, in part, due to the lack of evidence for clinical effectiveness of genetic tests. Recently, the use of whole-genome sequencing brought up an additional concern regarding how incidental findings should be communicated to the patient. On top of this, there is a very limited pace of translating genomic research findings into genomic medicine applications, which almost always lack the relevant economic evaluation evidence (Snyder et al., 2014).

Economic evaluation in genomic medicine has some specific requirements by nature. As such, certain approaches used in standard health economics, such as CEA or quality-adjusted life-years (QALYs), may not adequately capture the outcomes in genomic medicine and pharmacogenomics, whereas other elements related to the economic evaluation approach, such as stakeholder attitudes, societal perspective, and heterogeneity of cost data in different countries, are integral to the evaluation but complicate the economic evaluation.

The preceding paragraph argues that existing economic evaluation methods may not be adequate for the evaluation and technology assessment of genomic medicine interventions. In particular, it should be clarified which economic evaluation approach should be used for a particular intervention and/or technology to be adopted into routine clinical practice; this is a task that becomes more complicated as more advanced—and reciprocally more expensive—genomic tests, such as array-based genetic screening or next-generation sequencing methods, emerge. To this end, the battery of tools for health economists will be further enriched for them to provide their expertise to translate genomics data into the clinic (Van Rooij et al., 2012).

In brief, there are four main items that should be taken into consideration for economic evaluation in genomic medicine and health technology assessment, namely analysis costs, analysis method, outcome of analysis, and analysis effectiveness. These items are described in detail in the following paragraphs.

ANALYSIS COSTS

The costs of each medical intervention constitute the cornerstone of all economic evaluations. In contrast to cost data used in classic economic evaluation, in genomic medicine there is often uncertainty regarding which costs should be collected, the time that they should be collected, and how costs vary between different laboratories and health care systems. There are a variety of factors that lead to this problem. For example, in the United States, reimbursement for genetic tests has been based on a "stack" of procedure-based codes (e.g., DNA extraction and purification, amplification, sequencing, etc.). Different laboratories could use different procedures to generate results, meaning that the same genetic test can have markedly different costs and reimbursement due to the specific procedure that was used. Another issue is that a germline genetic test result does not change over the course of a person's lifetime and thus could be used multiple times for care decisions. An example of this is the pharmacogene *CYP2D6*, which is estimated to be involved in the oxidative metabolism of 25% of the drugs used in patient care (Samer et al., 2013). How would one perform a cost-effectiveness analysis for *CYP2D6* if the horizon is the patient's life span?

In genomic medicine, there are direct and indirect costs that should be taken into consideration for economically evaluating a genomics-related intervention (Table 7.1). This implies that health economists involved in such economic evaluation should ensure that all relevant costs are included in their analysis. Such costs would include patient recruitment (Giacomini et al., 2003; Rogowski et al., 2010), sample

Table 7.1 Main costs items involved in economic evaluation in Genomic Medicine

Direct costs

- Patient recruitment, sample collection (blood, saliva, buccal swabs, etc.)
- Nucleic acid isolation
- Genetic testing, including amplification and purification
- Nature of genomic variant tested (germline vs. somatic)
- Data analysis, including data storage and analysis using informatics solutions and genomic databases
- Frequency of data analysis (on the basis of novel genomics research findings)
- Accreditation, Quality Control of genetic testing platforms and assays
- Genetic counseling and results communication
- Post-testing actions (including treatment options, follow-up testing, monitoring, etc.)
- Training of personnel, health care professionals
- Infrastructure acquisition and maintenance

Indirect costs

- Patient's productivity loss
- Family costs (travel, accommodation, productivity loss)

collection (blood, saliva, buccal swabs, etc.) and delivery, genomic testing (Giacomini et al., 2003), genetic data analysis, which includes the appropriate genome informatics tools and databases (Feero et al., 2013), reporting of genetic test results to the interested parties, namely patients and their physicians, and genetic counseling (Veenstra et al., 2000; Payne, 2007; Faulkner et al., 2012; Singer and Watkins, 2012). Those costs should be also accompanied by the costs accrued by the medical interventions dictated by the genetic test results, such as costs of treatment (from which the costs from the adverse drug reactions and follow-up tests avoided due to the genetic testing should be deducted). The latter costs can be significant. For example, a targeted resequencing or a microarray-based approach, both of which are truly expensive tests on a stand-alone basis, to identify the genetic basis of a rare mental disorder could be cost-effective if the costs for follow-up testing can be avoided (Wordsworth et al., 2007). As noted in Chapter 6, in the case of chronic hepatitis C, even a small increment of improvement due to a genomic intervention can be cost-effective if the consequences of treatment failure (end-stage liver disease hepatocellular carcinoma) are very costly (Bock et al., 2014).

Another important cost item involves drug response and/or disease progression monitoring. In the case of warfarin treatment in which monitoring, for example, $CYP2C9$ genotyping, is relatively cheap (Veenstra et al., 2000), genomic interventions to individualize treatment may not be cost-effective. At this point, it must be noted that this might not be the case in developing countries, where genotyping and follow-up treatment costs may vary (Mitropoulou et al., 2015). As such, genomic tests that are more likely to be cost-effective are those for conditions for which monitoring is expensive and cumbersome (Veenstra et al., 2000) or when adverse events and treatment failures that could be prevented by the use of genomic testing are very expensive (Kauf et al., 2010; Bock et al., 2014). Unfortunately, there are not yet established guidelines and reimbursement rates for genomic-based interventions and testing because the genomic technologies either are too new for reimbursement rates to have been established, such as whole-genome sequencing, or are integrated as part of a cost item for an entire treatment modality, in which case costs may vary considerably depending on the genetic testing laboratory cost policy (Payne, 2009; Van Rooij et al., 2012; Singer and Watkins, 2012; Deverka et al., 2012; Malhotra et al., 2012). The latter differences may be the result of the use of either laboratory-developed assays versus commercially available and quality-certified genotyping kits (De Leon, 2009; Djalalov et al., 2011) or the variable prices among different countries (van den Akker-van Marle et al., 2006). In the latter case, it is noteworthy that genome-guided warfarin treatment may be cost-effective in developing countries (Mitropoulos et al., 2011) that do not have dedicated anticoagulation monitoring clinics but not cost-effective in developed countries (Mitropoulou et al., 2015), which makes it difficult if not impossible to generalize economic evaluation results between different health care systems.

The nature of the genomic variant being tested is also an important parameter that should be taken into consideration in economic evaluation studies pertaining to genomic medicine. Genomic variants may be either germline, in which case genetic testing should only be performed once during a patient's lifetime (De Leon, 2009), or acquired/somatic, which in some cases requires recurrent genetic analyses, like in the case of chronic lymphocytic leukemia (Knight et al., 2012). In addition, for new technologies such as next-generation sequencing, data analysis with its attendant costs may need to be repeated more than once because genomic data interpretation is frequently updated with new genomics research results, allowing for novel genotype—phenotype correlations on data re-evaluation (Gurwitz et al., 2009; Rogowski et al., 2010).

Apart from the direct costs indicated previously, there are also indirect costs such as the productivity costs (e.g., time lost from work) or the costs for a patient to seek the proper treatment, which reflects the time that will be lost from work. To return to the warfarin cost-effectiveness example, although the cost-effectiveness may be marginal, as noted by Meckley et al. (2010), from the patient perspective having two to three fewer INR measurements represents a significant reduction in life disruption that, if quantified, could impact the cost-effectiveness conclusion. Another example would be the use of a gene expression panel to stratify risk of recurrence in endocrine receptor—positive, node-negative breast cancer patients to determine which patients are less likely to benefit from adjunct chemotherapy based on a low recurrence risk (Vataire et al., 2012; Carlson and Roth, 2013). These patients could choose to forego chemotherapy with its attendant morbidity with a modest impact on cancer-related outcomes, supporting this as a cost-effective (or possibly even cost-saving) intervention (Rouzier et al., 2013).

Finally, education and training costs are additional cost items in genomic medicine economic evaluation studies (Reydon et al., 2012; Kampourakis et al., 2014), although these costs may be difficult to calculate and/or model accurately because of the complexity of genetic tests performed in a single instrument (e.g., real-time PCR or sequencer) or the economy of scale (multiple instruments working in parallel) that would contribute in reducing the overall analysis costs.

ANALYSIS METHOD

According to the few available economic evaluation studies in genomic medicine, it seems that no golden rule is currently available, contrary to the majority of economic evaluation studies of classical (nongenomic) interventions that take into consideration the direct impact of a medical intervention on the health care system and public health (Mette et al., 2012). The information provided by genomic analysis, particularly whole-genome sequencing performed very early in life, can have long-term implications that are not taken into consideration with classical economic evaluation studies. Consider identifying an α-synuclein gene variant leading to Parkinson disease in young asymptomatic patients. In this case, classical studies fail to estimate the long-term cumulative costs and effects for such a patient (Rogowski et al., 2007, 2009; Grosse et al., 2010).

Also, the timing of an economic evaluation study may be an equally important parameter because the genetic testing costs are rapidly decreasing, whereas their specificity and accuracy are steadily increasing (Conti et al., 2010; Goddard et al., 2012; Lin et al., 2012). This fact may lead to patient subgroup stratification with direct impact on individualized treatments, or the incorporation in the clinical practice of microarray-based genetic screening tests (e.g., AmpliChip [Roche Diagnostics], DMET+ [Affymetrix], etc.) that were previously used for research purposes only.

Cost-effectiveness of a specific genomic technology and/or a genome-based intervention can only be assessed in the context of a specific clinical application involving a certain patient subpopulation. As such, when performing economic evaluation for such a technology or intervention, it is of utmost importance to carefully identify and select the comparators. For example, one should assess both genetic and nongenetic testing in addition to genetic in conjunction with nongenetic testing (Sanderson et al., 2005; Hall et al., 2012); in the latter case, combining a cheap conventional screening approach with an expensive genetic test may prove to be cost-effective, like in the case of combining immunohistochemistry with DNA sequencing, respectively, for the identification of patients with Lynch syndrome among newly diagnosed patients with colorectal cancer (Mvundura et al., 2010). In the case of warfarin, the impact of nongenetic patient factors has at least as great an impact on warfarin dose as does the pharmacogenomics variants and, as such, they are included in warfarin dosing algorithms such as those found on warfarindosing.org. A study of warfarin dosing based on genomic factors alone compared with clinical factors alone would not be deemed safe, which is why all warfarin pharmacogenomics studies have compared a conventional dosing algorithm with a conventional plus genomic algorithm.

It is important to carefully evaluate the appropriate perspective from which the economic analysis is performed, because the results will vary depending on the stakeholder perspective that is chosen. Although most cost-effectiveness studies are done from the societal perspective, the utility of the results for a given stakeholder perspective may be limited based on the health care environment of the stakeholder.

Finally, economic evaluation studies should be restricted to genomic medicine interventions that are approved by regulatory bodies because the strength of scientific evidence about the genomic biomarker associated with the resulting clinical outcome(s) may directly impact the cost-effectiveness of a therapeutic intervention (e.g., the cost-effectiveness of Oncotype DX varies depending on whether it is being applied in breast or colon cancer; Alberts et al., 2013; Holt et al., 2013).

OUTCOME OF ANALYSIS

There are several methodological issues in economic evaluation studies in genomic medicine that arise when measuring the outcomes of a genomic intervention, such as the type of outcome measure to be used in the analysis, personal utility, and the importance of individual outcomes. The last of these is particularly relevant to genomic medicine because, although the probability of an individual outcome is considered in economic analyses, the predictive nature of genomic information provides more detailed information specific to individual patients. This has led some to conclude that new types of studies are needed to truly capture the impact of genomic medicine (Hu, 2012; Garraway, 2013). If true, then this will inevitably impact economic analyses as well.

In economic evaluation studies for genomic medicine interventions, preference-based outcome measures (e.g., the EuroQol five-dimensional [EQ-5D] questionnaire) should be used because they can be more easily and uniformly compared among different population subgroups by collecting data across a broad range of health-related quality-of-life domains.

However, QALYs measured using these measures do not reflect all possible health states applicable for genomic interventions (e.g., an asymptomatic patient with a predisposing genomic variant).

Genomic information, especially in the case of incidental findings, may offer significant benefits and, at the same time, harms for a person. Some genomic information lacks clinical utility (that is, there are no medical actions that can be taken that can prevent or treat the condition identified by the genomic test). This type of information can affect a person's well-being and decision-making (Foster et al., 2009; Roth et al., 2011; Garau et al., 2013), a concept referred to as personal utility. There are factors that have a positive impact on personal utility, such as prognostic or diagnostic information that explains a certain genotype–phenotype correlation, ends a diagnostic odyssey, or rationalizes a therapeutic intervention. Also, genomic information can be reassuring and can relieve anxiety (Asch et al., 1996; Caughey, 2005; Grosse et al., 2010), allowing patients and their relatives to better plan their lives and make lifestyle decisions, such as in the case of *APOE* E4 allele testing as a predisposing factor in Alzheimer's disease (Grosse et al., 2009, 2010; Payne et al., 2012). At the same time, there are factors that have a negative impact on personal utility. Anxiety may be increased if a genetic test result indicates that a patient is a nonresponder or is at increased risk for development of adverse reactions to a certain medical treatment, when no treatment alternatives exist (Conti et al., 2010; Khoury et al., 2011), or when mutations of unknown significance or other incidental findings are false-positives results (Jarrett and Mugford, 2006; Foster et al., 2009; Payne et al., 2012; Patrinos et al., 2013). In other words, if a patient is genomically diagnosed with a predisposition for development of Alzheimer's disease prior to the development of symptoms, this does not provide them with any clinical utility as traditionally defined because there are no effective treatments to prevent development of the disease. However, it considerably impacts their personal utility, sometimes in a positive way through adoption of a healthier lifestyle or purchase of long-term care insurance, but it can also lead to anxiety, despair, and pursuit of ineffective therapies that incur cost but provide no benefit (Roberts et al., 2011). Such result may also worsen family dynamics, lead to fear of discrimination and stigmatization (Grosse et al., 2008), and impact reproductive decisions (Jarrett and Mugford, 2006; Foster et al., 2009; Rogowski et al., 2010). False-negative genetic test results can also lead to false reassurance and encourage unhealthy behaviors, such as smoking and drinking (Grosse et al., 2008).

It is obvious that although personal utility is an important measure in this context, it is very difficult to incorporate into economic evaluations because we cannot accurately measure personal utility. Returning to the Alzheimer's disease example, some early work utilizing willingness to pay for genetic testing is being used to explore the valuation of personal utility (Kopits et al., 2011). Discussions among experts in genomics and economics are needed to develop new approaches that could potentially overcome these issues (Buchanan et al., 2013).

ANALYSIS EFFECTIVENESS

Economic evaluations also incorporate some measure of the effectiveness of the intervention being evaluated, such as the unpredictable behavior of patients and their physician. The complexity of genomic effectiveness data makes the task of measuring the effectiveness of the analysis a challenging one.

It is well-understood that the cost-effectiveness of a certain genomic intervention is dependent on how patients and their physicians use these interventions and how they respond to the genomic test results (e.g., whether patients change their lifestyle or alter their treatment modalities). As such, information regarding behavioral responses should be included into economic calculations to evaluate genomic technology and a genomic medicine intervention (Asch et al., 1996; Caughey, 2005; Rogowski et al., 2007; 2010; Conti et al., 2010). However, very little information regarding how to model behavioral attitudes is available (Conti et al., 2010). One cannot be sure whether all patients will want to have their genome, in part or as a whole, tested, whether they will they adjust their behavior to comply with the advice of a genetic counselor (Rogowski, 2009; Deverka et al., 2012), or whether physicians will take information from genomic tests into account when making clinical decisions (Conti et al., 2010). In some case, physicians confess that their genetics literacy is very limited to make informed decisions based on genetic test results (Mai et al., 2011, 2014). Some investigators such as those referenced in the Oncotype DX and Alzheimer's disease examples are using study designs that capture patient and physician choices in response to genomic information, thus generating data that can be used in subsequent economic analyses. These analyses will have much more relevance to real-world decision-making compared with studies that rely on untested assumptions about patient and physician behaviors.

The level of evidence linking complex phenotypes and genomic data to measurable health outcomes is generally very limited (Van Rooij et al., 2012; Guzauskas et al., 2013), which makes it a challenging task to incorporate the information provided by genomic technologies into economic evaluations in genomic medicine. For example, although there are more than 150 *CYP2D6* pharmacogene variants that can be used as pharmacogenomic biomarkers to more than 200 of the most commonly prescribed drugs in oncology, psychiatry, and other fields, it is mostly unclear which of these biomarkers are most clinically relevant (Van den Akker-van Marle et al., 2006; EGAPP, 2007). Also, the current performance of *CYP2D6* genotyping assays is highly variable, leading to uncertainty about the analytic validity of the testing (Fang et al., 2014). Other issues that may arise include the uncertainty of how to incorporate diagnostic characteristics into economic analyses, such as the analytic performance of commercially available versus laboratory genetic diagnostic tests, which may vary considerably (Garau et al., 2013) and for which there are limited published data.

CONCLUSIONS

From the previous paragraphs, it is obvious that economic evaluation in genomic medicine and health technology assessment has significant differences when compared with classical health economics and poses a number of challenges for health economists. To date, there are some solutions to these challenges that have already been proposed and could be incorporated in future economic evaluations in genomic medicine. The goal will be to introduce complex genomic effectiveness data into economic analyses as soon as next-generation sequencing is incorporated into clinical practice to facilitate informed decisions about implementation, resource allocation, and investment.

Also, it is of paramount importance to incorporate personal utility of genomic tests into the economic evaluation studies. In the United States and in other countries where the genetic tests are reimbursed, a flat cost-based scheme is used, leaving very little flexibility to consider personal utility of a genetic test. However, this may be more relevant for regulatory bodies rather than health care providers, whose aim is to improve public health and, as such, personal utility may not be so relevant (Grosse et al., 2008). Early efforts to value personal utility should be increased to develop effective solutions to this challenge.

Ideally, economic evaluation should be readily available and applicable to convince policy-makers to engage in nationwide research projects to use genomic approaches to relieve populations from major health burdens, such as β-thalassemia in countries in the Mediterranean belt and type II diabetes in Middle Eastern countries and China. Such genomic research findings would improve the quality of life of citizens and at the same time potentially reduce the overall health care expenditure. This way, economic evaluation would enable researchers to compare the expected benefits and cost-savings from implementing the results from this genomics research with the expected costs of undertaking this research, and it can highlight those cases for which new scientific evidence would be particularly valued by decision-makers. The pace of genomic technology innovation, especially in the dawn of the next-generation sequencing era, suggests that investing in expensive projects in the short-term may yield important results that would improve the quality of life and result in cost-savings in the longer-term. Also, such studies may strengthen the evidence of the effectiveness of specific genomic interventions by collecting additional data. Such studies may involve whole-genome sequencing analysis of large population groups from certain countries, such as the 100K genome project in the United Kingdom, in Saudi Arabia, and elsewhere.

Finally, new methods and models may be required to resolve some of the challenges outlined in the previous paragraphs and to improve the quality of economic evaluations, which should be the focus of future work in this field. Resolving these methodological issues will enable decision-makers and policymakers to make better informed adoption decisions, which may help accelerate the translation of pharmacogenomics and genomics research into genomic medical practice.

REFERENCES

Alberts, S.R., Yu, T., Behrens, R.J., et al., 2013. Real-world comparative economics of a 12-gene assay for prognosis in stage II colon cancer. J. Clin. Oncol. 31 (4), 391.

Asch, D.A., Hershey, J.C., Pauly, M.V., Patton, J.P., Jedrziewski, M.K., Mennuti, M.T., 1996. Genetic screening for reproductive planning: methodological and conceptual issues in policy analysis. Am. J. Public Health 86 (5), 684–690.

Bock, J.A., Fairley, K.J., Smith, R.E., Maeng, D.D., Pitcavage, J.M., Inverso, N.A., et al., 2014. Cost-effectiveness of IL28 genotype-guided protease inhibitor triple therapy versus standard of care treatment in patients with hepatitis C genotypes 2 or 3 infection. Public Health Genomics 17, 306–319.

Buchanan, J., Wordsworth, S., Schuh, A., 2013. Issues surrounding the health economic evaluation of genomic technologies. Pharmacogenomics 14 (15), 1833–1847.

Carlson, J.J., Roth, J.A., 2013. The impact of the Oncotype Dx breast cancer assay in clinical practice: a systematic review and meta-analysis. Breast Cancer Res. Treat. 141 (1), 13–22.

Caughey, A.B., 2005. Cost-effectiveness analysis of prenatal diagnosis: methodological issues and concerns. Gynecol. Obstet. Invest. 60 (1), 11–18.

Conti, R., Veenstra, D.L., Armstrong, K., Lesko, L.J., Grosse, S.D., 2010. Personalized medicine and genomics: challenges and opportunities in assessing effectiveness, cost-effectiveness, and future research priorities. Med. Decis. Making 30 (3), 328–340.

De Leon, J., 2009. Pharmacogenomics: the promise of personalized medicine for CNS disorders. Neuropsychopharmacology 34 (1), 159–172.

Deverka, P.A., Schully, S.D., Ishibe, N., Carlson, J.J., Freedman, A., Goddard, K.A., et al., 2012. Stakeholder assessment of the evidence for cancer genomic tests: insights from three case studies. Genet. Med. 14 (7), 656–662.

Djalalov, S., Musa, Z., Mendelson, M., Siminovitch, K., Hoch, J., 2011. A review of economic evaluations of genetic testing services and interventions (2004–2009). Genet. Med. 13 (2), 89–94.

Fang, H., Liu, X., Ramírez, J., Choudhury, N., Kubo, M., Im, H.K., et al., 2014. Establishment of CYP2D6 reference samples by multiple validated genotyping platforms. Pharmacogenomics J. 14, 564–572.

Faulkner, E., Annemans, L., Garrison, L., Helfand, M., Holtorf, A.P., Hornberger, J., et al., 2012. Challenges in the development and reimbursement of personalized medicine—payer and manufacturer perspectives and implications for health economics and outcomes research: a report of the ISPOR personalized medicine special interest group. Value Health 15 (8), 1162–1171.

Feero, W., Wicklund, C., Veenstra, D.L., 2013. The economics of genomic medicine: insights from the iom roundtable on translating genomic-based research for health. JAMA 309 (12), 1235–1236.

Flore, L.A., Milunsky, J.M., 2012. Updates in the genetic evaluation of the child with global developmental delay or intellectual disability. Semin. Pediatr. Neurol. 19 (4), 173–180.

Foster, M.W., Mulvihill, J.J., Sharp, R.R., 2009. Evaluating the utility of personal genomic information. Genet. Med. 11 (8), 570–574.

Garau, M., Towse, A., Garrison, L., Housman, L., Ossa, D., 2013. Can and should value based pricing be applied to molecular diagnostics? Pers. Med. 10 (1), 61–72.

Garraway, L.A., 2013. Genomics-driven oncology: framework for an emerging paradigm. J. Clin. Oncol. 31 (15), 1806–1814.

Giacomini, M., Miller, F., O'Brien, B.J., 2003. Economic considerations for health insurance coverage of emerging genetic tests. Community Genet. 6 (2), 61–73.

Goddard, K.A., Knaus, W.A., Whitlock, E., Lyman, G.H., Feigelson, H.S., Schully, S.D., et al., 2012. Building the evidence base for decision making in cancer genomic medicine using comparative effectiveness research. Genet. Med. 14 (7), 633–642.

Gordon, A.S., Smith, J.D., Xiang, Q., Metzker, M.L., Gibbs, R.A., Mardis, E.R., et al., PGRNseq: a new sequencing-based platform for high-throughput pharmacogenomic implementation and discovery. Abstract presented at the 62nd Annual Meeting of the American Society of Human Genetics: November 6–8, 2012; San Francisco, CA. Program #244. <http://www.ashg.org/2012meeting/abstracts/fulltext/f120122669.htm>.

Grosse, S.D., Kalman, L., Khoury, M.J., 2010. Evaluation of the validity and utility of genetic testing for rare diseases. Adv. Exp. Med. Biol. 686, 115–131.

Grosse, S.D., McBride, C.M., Evans, J.P., Khoury, M.J., 2009. Personal utility and genomic information: look before you leap. Genet. Med. 11 (8), 575–576.

Grosse, S.D., Rogowski, W.H., Ross, L.F., Cornel, M.C., Dondorp, W.J., Khoury, M.J., 2010. Population screening for genetic disorders in the 21st century: evidence, economics, and ethics. Public Health Genomics 13 (2), 106—115.

Grosse, S.D., Wordsworth, S., Payne, K., 2008. Economic methods for valuing the outcomes of genetic testing: beyond cost-effectiveness analysis. Genet. Med. 10 (9), 648—654.

Gurwitz, D., Zika, E., Hopkins, M.M., Gaisser, S., Ibarreta, D., 2009. Pharmacogenetics in Europe: barriers and opportunities. Public Health Genomics 12 (3), 134—141.

Guzauskas, G.F., Garrison, L.P., Stock, J., Au, S., Doyle, D.L., Veenstra, D.L., 2013. Stakeholder perspectives on decision-analytic modeling frameworks to assess genetic services policy. Genet. Med. 15 (1), 84—87.

Hall, P.S., McCabe, C., Stein, R.C., Cameron, D., 2012. Economic evaluation of genomic test-directed chemotherapy for early-stage lymph node-positive breast cancer. J. Natl. Cancer Inst. 104 (1), 56—66.

Holt, S., Bertelli, G., Humphreys, I., Valentine, W., Durrani, S., Pudney, D., et al., 2013. A decision impact, decision conflict and economic assessment of routine Oncotype DX testing of 146 women with node-negative or pNlmi, ER-positive breast cancer in the UK. Br. J. Cancer 108 (11), 2250—2258.

Hu, V.W., 2012. Subphenotype-dependent disease markers for diagnosis and personalized treatment of autism spectrum disorders. Dis. Markers. 33 (5), 277—288.

Jarrett, J., Mugford, M., 2006. Genetic health technology and economic evaluation: a critical review. Appl. Health Econ. Health Policy. 5 (1), 27—35.

Kampourakis, K., Vayena, E., Mitropoulou, C., Borg, J., van Schaik, R.H., Cooper, D.N., et al., 2014. Key challenges for next generation pharmacogenomics. EMBO Rep. 15 (5), 472—476.

Kauf, T.L., Farkouh, R.A., Earnshaw, S.R., Watson, M.E., Maroudas, P., Chambers, M.G., 2010. Economic efficiency of genetic screening to inform the use of abacavir sulfate in the treatment of HIV. Pharmacoeconomics 28 (11), 1025—1039.

Khoury, M.J., Gwinn, M., Dotson, W.D., Bowen, M.S., 2011. Is there a need for PGxceptionalism? Genet. Med. 13 (10), 866—867.

Knight, S.J.L., Yau, C., Clifford, R., Timbs, A.T., Sadighi Akha, E., Dréau, H.M., et al., 2012. Quantification of subclonal distributions of recurrent genomic aberrations in paired pre-treatment and relapse samples from patients with B-cell chronic lymphocytic leukemia. Leukemia 26 (7), 1564—1575.

Kopits, I.M., Chen, C., Roberts, J.S., Uhlmann, W., Green, R.C., 2011. Willingness to pay for genetic testing for Alzheimer's disease: a measure of personal utility. Genet. Test. Mol. Biomarkers 15 (12), 871—875.

Laduca, H., Stuenkel, A.J., Dolinsky, J.S., Keiles, S., Tandy, S., Pesaran, T., et al., 2014. Utilization of multigene panels in hereditary cancer predisposition testing: analysis of more than 2,000 patients. Genet Med. 16, 830—837.

Lin, X., Tang, W., Ahmad, S., et al., 2012. Applications of targeted gene capture and next-generation sequencing technologies in studies of human deafness and other genetic disabilities. Hear. Res. 288 (1—2), 67—76.

Mai, Y., Koromila, T., Sagia, A., Cooper, D.N., Vlachopoulos, G., Lagoumintzis, G., et al., 2011. A critical view of the general public's awareness and physicians' opinion of the trends and potential pitfalls of genetic testing in Greece. Pers. Med. 8 (5), 551—561.

Mai, Y., Mitropoulou, C., Papadopoulou, X.E., Vozikis, A., Cooper, D.N., van Schaik, R.H., et al., 2014. Critical appraisal of the views of healthcare professionals with respect to pharmacogenomics and personalized medicine in Greece. Pers. Med. 11 (1), 15—26.

Malhotra, A.K., Zhang, J.P., Lencz, T., 2012. Pharmacogenetics in psychiatry: translating research into clinical practice. Mol. Psychiatr 17 (8), 760—769.

Mardis, E.R., 2008. The impact of next-generation sequencing technology on genetics. Trends Genet. 24 (3), 133—141.

Meckley, L.M., Gudgeon, J.M., Anderson, J.L., Williams, M.S., Veenstra, D.L.A., 2010. Policy model to evaluate the benefits, risks and costs of warfarin pharmacogenomic testing. Pharmacoeconomics 28 (1), 61–74.

Mette, L., Mitropoulos, K., Vozikis, A., Patrinos, G.P., 2012. Pharmacogenomics and public health: implementing populationalized medicine. Pharmacogenomics 13 (7), 803–813.

Mitropoulos, K., Johnson, L., Vozikis, A., Patrinos, G.P., 2011. Relevance of pharmacogenomics for developing countries in Europe. Drug Metabol. Drug Interact. 26 (4), 143–146.

Mitropoulou, C., Fragoulakis, V., Bozina, N., Vozikis, A., Supe, S., Bozina, T., et al., 2015. Economic evaluation for pharmacogenomic-guided warfarin treatment for elderly croatian patients with atrial fibrillation. Pharmacogenomics. (in press).

Mizzi, C., Mitropoulou, C., Mitropoulos, K., Peters, B., Agarwal, M.R., van Schaik, R.H., et al., 2014. Personalized pharmacogenomics profiling using whole genome sequencing. Pharmacogenomics 15, 1223–1234.

Mvundura, M., Grosse, S.D., Hampel, H., Palomaki, G.E., 2010. The cost-effectiveness of genetic testing strategies for lynch syndrome among newly diagnosed patients with colorectal cancer. Genet. Med. 12 (2), 93–104.

Patrinos, G.P., Baker, D.J., Al-Mulla, F., Vasiliou, V., Cooper, D.N., 2013. Genetic tests obtainable through pharmacies: the good, the bad and the ugly. Hum. Genomics 7 (1), 17.

Payne, K., 2007. Towards an economic evidence base for pharmacogenetics: consideration of outcomes is key. Pharmacogenomics 9 (1), 1–4.

Payne, K., 2009. Fish and chips all round? Regulation of DNA-based genetic diagnostics. Health Econ. 18 (11), 1233–1236.

Payne, K., McAllister, M., Davies, L.M., 2012. Valuing the economic benefits of complex interventions: when maximising health is not sufficient. Health Econ. 22 (3), 258–271.

Recommendations from the EGAPP Working Group: testing for cytochrome P450, 2007. Polymorphisms in adults with nonpsychotic depression treated with selective serotonin reuptake inhibitors. Evaluation of genomic applications in practice and prevention (EGAPP) working group. Genet Med. 9 (12), 819–825.

Reydon, T.A., Kampourakis, K., Patrinos, G.P., 2012. Genetics, genomics and society: the responsibilities of scientists for science communication and education. Pers. Med. 9 (6), 633–643.

Roberts, J.S., Christensen, K.D., Green, R.C., 2011. Using Alzheimer's disease as a model for genetic risk disclosure: implications for personal genomics. Clin. Genet. 80 (5), 407–414.

Rogowski, W.H., 2007. Current impact of gene technology on healthcare. A map of economic assessments. Health Policy 80 (2), 340–357.

Rogowski, W.H., 2009. The cost-effectiveness of screening for hereditary hemochromatosis in Germany: a remodeling study. Med. Decis. Making 29 (2), 224–238.

Rogowski, W.H., Grosse, S.D., John, J., Kääriäinen, H., Kent, A., Kristofferson, U., et al., 2010. Points to consider in assessing and appraising predictive genetic tests. J. Community Genet. 1 (4), 185–194.

Rogowski, W.H., Grosse, S.D., Khoury, M.J., 2009. Challenges of translating genetic tests into clinical and public health practice. Nat. Rev. Genet. 10 (7), 489–495.

Roth, J.A., Garrison Jr., L.P., Burke, W., Ramsey, S.D., Carlson, R., Veenstra, D.L., 2011. Stakeholder perspectives on a risk-benefit framework for genetic testing. Public Health Genomics 14 (2), 59–67.

Rouzier, R., Pronzato, P., Chéreau, E., Carlson, J., Hunt, B., Valentine, W.J., 2013. Multigene assays and molecular markers in breast cancer: systematic review of health economic analyses. Breast Cancer Res. Treat. 139 (3), 621−637.

Samer, C.F., Lorenzini, K.I., Rollason, V., Daali, Y., Desmeules, J.A., 2013. Applications of CYP450 testing in the clinical setting. Mol. Diagn. Ther. 17 (3), 165−184.

Sanderson, S., Zimmern, R., Kroese, M., Higgins, J., Patch, C., Emery, J., 2005. How can the evaluation of genetic tests be enhanced? Lessons learned from the ACCE framework and evaluating genetic tests in the United Kingdom. Genet. Med. 7 (7), 495−500.

Singer, D.R.J., Watkins, J., 2012. Using companion and coupled diagnostics within strategy to personalize targeted medicines. Pers. Med. 9 (7), 751−761.

Snyder, S., Mitropoulou, C., Patrinos, G.P., Williams, M.S., 2014. Economic evaluation of pharmacogenomics: a value-based approach to pragmatic decision-making in the face of complexity. Public Health Genomics 17, 256−264.

van den Akker-van Marle, M.E., Gurwitz, D., Detmar, S.B., Enzing, C.M., Hopkins, M.M., Gutierrez de Mesa, E., et al., 2006. Cost-effectiveness of pharmacogenomics in clinical practice: a case study of thiopurine methyltransferase genotyping in acute lymphoblastic leukemia in Europe. Pharmacogenomics 7 (5), 783−792.

Van Rooij, T., Wilson, D.M., Marsh, S., 2012. Personalized medicine policy challenges: measuring clinical utility at point of care. Expert Rev. Pharmacoecon. Outcomes Res. 12 (3), 289−295.

Vataire, A.L., Laas, E., Aballea, S., Gligorov, J., Rouzier, R., Chereau, E., 2012. Cost-effectiveness of a chemotherapy predictive test. Bull. Cancer 99 (10), 907−914.

Veenstra, D.L., Higashi, M.K., Phillips, K.A., 2000. Assessing the cost-effectiveness of pharmacogenomics. AAPS PharmSci. 2 (3), E29.

Wordsworth, S., Buchanan, J., Regan, R., Davison, V., Smith, K., Dyer, S., et al., 2007. Diagnosing idiopathic learning disability: a cost-effectiveness analysis of microarray technology in the national health service of the United Kingdom. Genomic Med. 1 (1−2), 35−45.

CHAPTER 8

A New Methodological Approach for Cost-Effectiveness Analysis in Genomic Medicine

INTRODUCTION

As mentioned in Chapters 4 and 5, economic evaluation compares the costs and health effects of an intervention aiming to maximize health benefits from available resources (Cartwright, 1999). So far, for innovative interventions, the most commonly used formal investigation is implemented via the determination of incremental cost-effectiveness ratio (ICER; O'Brien and Briggs, 2002; see Chapter 4). ICER is defined as the ratio of incremental cost (ΔC) of an intervention divided by the incremental effectiveness (ΔE), measured either as additional life-years (LY) or additional quality-adjusted life-years (QALYs). Based on the classical decision-making approach of the cost-effectiveness analysis, *ICER* is compared with a fixed amount that a policymaker is willing to pay (λ) for an additional LY or QALY. In this methodology context, if ICER is below λ, the new intervention meets the basic criterion for reimbursement by the payers (McCabe et al., 2008). In practice, this methodology assumes that the policymakers are willing to pay this amount, regardless of the amount of QALYs "purchased," and in that way are ignoring the budget constraint. However, certain limitations must be considered in this approach because: (i) there is no clear linkage between λ and the budget affordability of health services (Towse, 2009) and (ii) it assumes, silently, an instant adjustment of budget to absorb the cost of innovation (Gafni and Birch, 2006). Thus, the adoption of the existing process has led to the adoption of costly decisions (Sendi et al., 2002; Gyrd-Hansen, 2005; Birch and Gafni, 2006) and consequently increased expenditures for health care systems (Gafni and Birch, 2006). Even if the budget does not face hard constraints, it is reasonable to assume that there is an upper budget bound, especially today in the presence of a global economic crisis. In addition,

λ has been traditionally determined one-dimensionally by existing theory, without any consideration of the size of actual ΔE. Hence, λ has been considered insensitive to the actual ΔE as a criterion for reimbursement. In other words, in this methodology context, technology with a big influence on human life for a single person (namely with high ΔE in terms of QALYs compared with the standard therapy) would be reimbursed proportionally equally (e.g., up to €50,000 per QALY gained) with those technologies that would provide only a modest incremental health benefit (small ΔE). A similar concept applies to those technologies that are incrementally less effective but also less expensive compared with the standard ones (negative ΔE). The theory assumes that on the conditions of constrained resources and a fixed willingness to accept (wta), cost-saving interventions could be reimbursed by permitting more efficient resource reallocation across health care sectors (O'Brien et al., 2002; Nelson et al., 2009). This argument definitely imposes the need for accounting the cost of new interventions by the policymakers. However, there might be a lowest accepted bound difference between the standard and a new intervention because there are moral issues that prevent the adoption of a new, although far less effective, technology.

Several concerns have been raised regarding the methodological limitation of decision-making in the cost-effectiveness analysis, and there might be a need for an alternative approach (Gafni, 1998; Donaldson et al., 2002a,b; Barton et al., 2008; Eckermann et al., 2008; Whitehead and Ali, 2010). A comprehensive decision-making process should not be insensitive to the real impact on patients or to the payer's budget. Previous attempts to improve the decision-making process focused mainly on the problem of maximization of QALYs across health care sectors given a total health budget and using either linear or integer programming, depending on the model assumptions (perfect divisibility, deterministic or stochastic approach for outcomes, etc.; Weinstein and Zeckhauser, 1973; Stinnett and Paltiel, 1996). In such a case, some limitations must be addressed (Sendi et al., 2002; Birch and Gafni, 2006). For instance, the actual determination of λ in this kind of model requires the explicit knowledge of societal utility function, which has to be maximized, and the explicit knowledge of practical importance for QALYs, which come from different therapeutic areas (Wailoo et al., 2009). Of course in theory a "QALY is a QALY," but in practical applications it is somewhat difficult to compare an intervention in cardiology with one in oncology or rheumatology or other field to extract a common measure for comparisons among them. However, only if these prerequisites were fully satisfied would the true value of λ or the desired mix of budget across several disease-specific areas have been estimated. Mathematical programming can incorporate some restrictions referred to as ethical or technical issues (minimum budget spent on a specific disease, divisibility, etc.), but an amount of exogenous information for the pattern of budget allocation must also be provided as an input to solve the maximization problem. These are some crucial problems that face practical research today. In this chapter, we propose a new simple way of thinking and an alternative methodological approach for decision-making that takes into account the budget constraint and relaxes the constant return on scale assumption of λ.

NEW METHODOLOGICAL FRAMEWORK OF THE DECISION-MAKING PROCESS IN GENOMIC MEDICINE

Let us introduce or repeat some simple notations:

S = standard/conventional intervention; T = new intervention; E = effectiveness; C = cost

B_t = the total available budget the state is willing to pay for a patient's lifespan (t) (measured in a monetary unit such as € or $)

C_t = the total cost per patient of the new intervention

C_s = the total cost per patient of the standard/conventional intervention

$C(T)_{py}$ = the average cost per patient per year of the new intervention

$C(S)_{py}$ = the average cost per patient per year of the standard/conventional intervention

E_t = the effectiveness of the new intervention (measured in QALYs)

E_s = the effectiveness of the standard/conventional intervention (measured in QALYs)

C_s = the total cost per patient of the standard/conventional intervention

ΔE = the difference in the effectiveness between the standard/conventional and the new intervention

ΔE_{max} = the maximum expected difference in effectiveness between the standard/conventional and the new intervention that could be additionally reimbursed given the budget

ΔC = the difference in the total cost between the standard/conventional and the new intervention per patient

Given that we are referring to the "standard/conventional" intervention, which has already been adopted from the health care system, it must be obvious that Eq. (8.1) must be holding true:

$$B_t = C(S)_{py} E_s \tag{8.1}$$

For instance, if the state is willing to pay €10,000 for a patient during the remaining lifetime (which is expected to be 2 years), the mean cost per year for this patient that could be covered is limited to €5,000 per year. Figure 8.1 depicts the relation between the average cost per patient per year and the effectiveness of the standard intervention via the rational function Eq. (8.1).

An increased effectiveness in a given budget B_t from the substitution of standard intervention with a new intervention with equal total cost must be followed by a decrease in the average cost per patient per year and vice versa. The area under the curve is equal for each point across the curve and expresses the total cost per patient. When a new and more

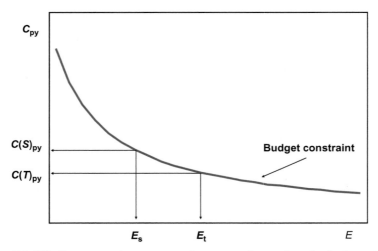

Figure 8.1 *Effectiveness and average cost per year for a given budget constraint.* C_{py}: average cost per patient per year; $C(S)_{py}$: average cost per patient per year for the standard intervention; $C(T)_{py}$: average cost per patient per year for the new intervention; budget constraint refers to the available budget per patient for his/her life span; E_s: effectiveness of the standard intervention; E_t: effectiveness of the new intervention.

costly intervention is introduced in the market, it must satisfy the following equation:

$$C(T)_{py}E_t \geq C(S)_{py}E_s \tag{8.2}$$

The reader must keep in mind that Eq. (8.2) is satisfied even if $C(T)_{py}$
is equal, lower, or higher than $C(S)_{py}$.

For instance, let us assume that the total cost of a new intervention has
been determined at 20,000 against 10,000 for the standard one. So, we have:
$C(T)_{py}E_t = 20,000$ and $C(S)_{py}E_s = 10,000$. If $E_t = 2$ LYs and $E_s = 1$ LY,
then $C(S)_{py} = C(T)_{py}$; if $E_t = 3$ and $E_s = 1$ LY, then $C(S)_{py} > C(T)_{py}$,
and so on.

This is depicted in Figures 8.2 and 8.3. Figure 8.2 depicts the case
where the most expensive and cost-infeasible interventions due to a specific
budget have equal, lower, or higher average costs per patient per year.
Figure 8.3 depicts the budget constraint expansion required for affording
new—more costly—interventions. Every single curve that is more distant
from the axes reveals an increased expenditure or a higher budget.

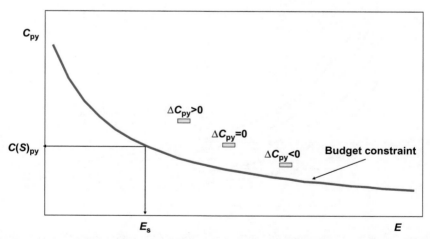

Figure 8.2 *Cost-infeasible interventions with lower, equal, and higher average cost
per year.* C_{py}: average cost per patient per year; $C(S)_{py}$: average cost per patient per
year for the standard intervention; ΔC_{py}: the difference in the average cost per year
per patient between the standard and new intervention; budget constraint refers to
the available budget per patient for his/her life span; E_s: effectiveness of the standard
intervention.

In a general form, the mean cost per patient per year for the new intervention could be expressed as:

$$C(T)_{py} = \frac{C_t}{E_t} \quad \text{or}$$

$$C(T)_{py} = \frac{C_t}{E_s + \Delta E} \quad \text{or}$$

$$C(T)_{py} = \frac{C_s + \Delta C}{E_s + \Delta E} \quad \text{or}$$

$$C(T)_{py} = \frac{C(S)_{py}E_s + \Delta C}{E_s + \Delta E}$$

(8.3)

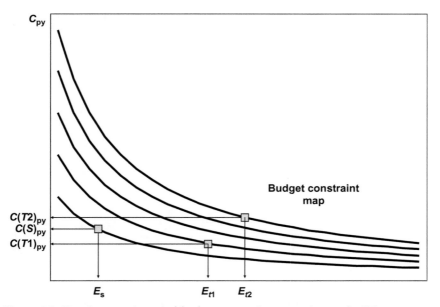

Figure 8.3 *New interventions and budget constraint expansion path.* $C(S)_{py}$: average cost per patient per year for the standard intervention; $C(T1)_{py}$: average cost per patient per year for the new T_1 intervention; $C(T2)_{py}$: average cost per patient per year for the new T_2 intervention; budget constraint refers to the available budget per patient for his/her life span; E_s: effectiveness of the standard intervention; E_{t1}: effectiveness of the new T_1 intervention; E_{t2}: effectiveness of the new T_2 intervention.

Let us assume that the threshold for a QALY is unknown but equal to λ. In addition to that, we know that $\Delta C = \lambda \, \Delta E$. Thus, we have:

$$C(T)_{py} = \frac{C(S)_{py}E_s + \lambda \Delta E}{E_s + \Delta E} \qquad (8.4)$$

From Eq. (8.4) it must be obvious that we have connected an incremental quantity (λ) with an average one ($C(T)_{py}$).

Thus, we can determine the expansion path of the average cost per patient per year when a new, more costly intervention is introduced as a function of ΔE, taking into account the (fixed) incremental threshold of λ. If $C(S)_{py} = \lambda$, then the graphical representation of $C(T)_{py}$ and E_t is a straight line. If $C(S)_{py} > \lambda$ ($C(S)_{py} < \lambda$), then the graphical representation is a downward (upward) curve with an asymptotic plateau of λ. Figure 8.4 depicts, for illustrative purposes, the case where $C(S)_{py} < \lambda$ Figure 8.4 depicts that the threshold approach leads to decisions that result in increased expenditures and thus raises concerns about the funding sustainability. Figure 8.5 shows the opposite case. Given a fixed budget constraint that has been exhausted, λ has a decreasing trend, is not constant across

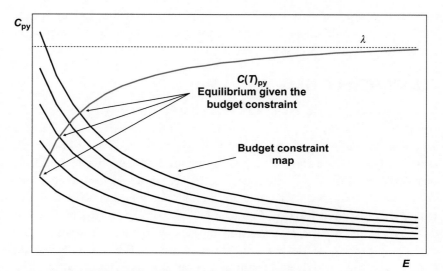

Figure 8.4 *Average cost and expanded budget constraint given a fixed* λ. C_{py}: average cost per patient per year; λ: willingness to pay; $C(T)_{py}$: average cost per patient per year for the new intervention; budget constraint refers to the available budget per patient for his/her life span; E: effectiveness.

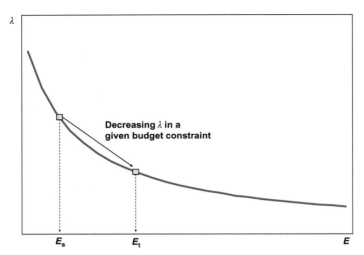

Figure 8.5 *Willingness to pay as a function of a fixed budget constraint and effectiveness.* λ: willingness to pay; E: effectiveness; E_s: effectiveness of the standard/conventional intervention; E_t: effectiveness of the new intervention.

ΔE, and becomes asymptotically zero. Thus, λ tends to be zero in the case of—potential—new interventions with high ΔE, because the society is unable to afford the incremental cost of innovation. If the ΔE tends to be small, then λ could be high; nonetheless, society probably is unwilling to pay a high incremental cost for a small ΔE.

RELAXATION OF FIXED λ

In this paragraph, we go a step further concerning λ. The presented approach is sensitive to the actual incremental effectiveness of a new intervention. Let us assume that the budget B_{2t} is fixed. The model assumes there is a maximum expected incremental effectiveness (ΔE_{max}) that could be additionally reimbursed. Hence, we could introduce a non-linear S-shape function that determines λ as a function of actual ΔE, where $\Delta E \geq 0$. As shown in Figure 8.6, the willingness to pay threshold belongs to the S-shape family, ensuring that λ is different across ΔE in every single point of the cost-effectiveness plane. For a very small ΔE (i.e., a "me too" drug), and without any substantial differentiation point from existing interventions, λ is expected to be low. If ΔE is fair or small, then a higher λ with an increased marginal utility for the additional QALYs can be expected. This is in accordance with economic theory

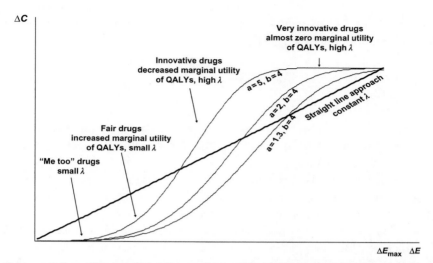

Figure 8.6 *A nonlinear S-shape function that determines the willingness to pay as a function of actual* Δ**E**. ΔE: difference in the effectiveness between the standard/conventional and the new intervention; ΔC: difference in the total cost between the standard/conventional and new intervention per patient; ΔC_{max}: available budget (per patient) is willing to afford the budget holder to capture the maximum expected effectiveness (ΔE_{max}); a and b are parameters of the equation:

$$\lambda = \frac{1 - e^{(-\alpha\Delta E^b)}}{\Delta E}\Delta C_{max}$$

and practical evidence because a small increase in life expectancy has an increased marginal utility (Kvamme et al., 2010) and the excess budget has a decreased marginal utility (Gafni and Birch, 2006). If a new intervention is very innovative in terms of ΔE, then λ must be diminishing because of diminishing marginal utility of QALYs and increased marginal utility of limited budget (Gafni and Birch, 2006).

In mathematical terms, relaxing the assumption of constant λ and expressing it as a function of ΔE, $\lambda = f(\Delta E)$, Eq. (8.4) becomes:

$$C(T)_{py} = \frac{C(S)_{py}E_s + \lambda(\Delta E)}{E_s + \Delta E} \tag{8.5}$$

Thus, the path of average cost, the budget constraint map, and the path of λ changes depend on the form of function of the willingness to pay threshold. Let us assume that λ has a flexible form derived from the Weibull family (Yin et al., 2003):

$$\lambda = \frac{1 - e^{(-a\Delta E^b)}}{\Delta E} \Delta C_{max} \qquad (8.6)$$

where:

ΔC_{max} = The maximum available budget (per patient) the budget holder is willing to afford to capture the maximum expected difference in effectiveness (ΔE_{max}).

ΔE_{max} has been assumed and a and b are constants. In such a case, the straight line converts to a line segment (bounded by 0 to the left and ΔE_{max} to the right). The new model partly incorporates the classical model because we could find appropriate values for the parameters a and b, which produce an approximately straight line segment or, to postulate it, a line in the cost-effectiveness plane. It is paramount to consider that in this model the budget upper limit is predetermined, and Figure 8.6 reveals the path of willingness to pay as we get closer to the budget constraint.

A SIMPLE INDICATIVE EXAMPLE

Let us assume that the survival (measured in QALYs) of standard intervention for a specific disease is estimated at 1 year ($E_s = 1$) and the total cost until death is €10,000 ($E_s = €10,000$). To keep it simple, let us assume that the maximum amount that a budget holder is willing to pay is €50,000 per patient ($\Delta C_{max} = €50,000$ per patient) if, and only if, 1 more year is captured for each patient ($\Delta E_{max} = 1$). The function that describes the willingness to pay path for $a = 5$ and $b = 4$ is:

$$\lambda = \frac{1 - e^{(-5\Delta E^4)}}{\Delta E} 50,000$$

assuming that a and b are parameters that have been estimated from real data. If a new intervention has a ΔE that is higher than expected (i.e., $\Delta E = 1.5$), then λ is estimated as $\lambda = \Delta C_{max}/\Delta E$ or $\lambda = 50,000/1.5 = €33,000$ per LY. In that case, the manufacturer must accept this λ due to budget constraints of the health care system and the inability of budget holders to absorb the cost of such an innovative (in term of ΔE) intervention. In the extreme scenario where the actual ΔE was even higher, λ tends to be zero, as depicted in Figure 8.5, but the benefit of new intervention disperses in the society on behalf of the patients. In the usual case where ΔE is less than 1 year, λ is estimated by Eq. (8.6).

For instance, let us assume an innovative technology with $E_t = 1.8$ (or $\Delta E = 0.8$) and $C_t = 43{,}000$. Thus, Eq. (8.6) estimates $\lambda = 54{,}438$. ICER is then estimated as ICER $= \Delta C / \Delta E$ or ICER $= 33{,}000/0.8 = 41{,}250$ per QALY. In this case, society is able to afford the new intervention and is willing to reimburse it because the ICER is lower than society's willingness to pay for this survival benefit of $\Delta E = 0.8$. It is noteworthy that λ was higher than $\Delta C_{max} = 50{,}000$ in $E_t = 1.8$ and the budget criteria were met. Thus, the model incorporates the budget criteria, the preferences, and the cost-effectiveness criteria at the same time.

AN APPLICATION OF THE NEW MODEL IN GENOMIC MEDICINE

To test the aforementioned concept using a genomic medicine scenario, we consider an economic evaluation analysis comparing pharmacogenomic (PGx)–guided treatment with conventional nonpharmacogenomic (N-PGx)-guided treatment. For the purpose of our study, we performed an economic evaluation of a pharmacogenomic-guided warfarin treatment in elderly patients with atrial fibrillation, considering that both PGX and N-PGx patients groups were homogeneous and stratified according to sex and age and then randomly assigned to the two different treatment arms.

In the PGx patient group, genotyping analysis included $CYP2C9^*2,^*3$, and $VKORC1$-1173C $>$ T pharmacogenomic biomarkers (see Chapter 3) prior to the prescription of warfarin, and the prescribed warfarin dose was calculated based on the genotype and the algorithm published in the nonprofit website http://www.WarfarinDosing.org. Subsequently, the doses were adjusted depending on the measured INR values. In patients with the "wild-type" (normal) $CYP2C9^*1/^*1$ and $VKORC1$-1173 CC genotypes, we applied a double dose of estimated doses during the first 2 days of treatment because it is known that these alleles are the slowest to achieve the target INR. After that, the doses were adjusted depending on the INR measurement. In the N-PGx control group consisting of the same number of patients with the same indication for anticoagulation and with the same criteria for entry in the study, warfarin was introduced by a fixed dose according to the standard criteria without pharmacogenomic analysis and doses were then adjusted depending on the INR values.

Based on the aforementioned available data and international literature, we constructed a pharmacoeconomic model to compare, from an
economic point of view, PGx-guided and N-PGx warfarin therapy in the
treatment of elderly patients with atrial fibrillation in a 1-year time
period. Our pharmacoeconomic model was a decision tree populated
with cost data, in line with current treatment guidelines on patient management, outcomes, and economic consequences. The model structure
was identical for each of the two assessed strategies.

In short, results showed that the total cost per patient was estimated to
be more than twofold higher for the PGx group (€540) compared with
the N-PGx control group (€230). This cost difference was due to the
genotyping costs. In terms of QALYs gained, the true difference in
QALYs was estimated at 0.01 in favor of the PGx group. The ICER of
the PGx versus the N-PGx control groups was estimated at ($\Delta C/
\Delta E = $€310/0.01) €31,000/QALY. Also, the results from the probabilistic
sensitivity analysis indicate that the treatment for the PGx group was
more expensive but also more effective than that for the N-PGx control
group.

Nonetheless, the cost-effectiveness of PGx is a subjective assessment
and depends on the willingness to pay per QALY gained with it. A convenient way of illustrating the results is the cost-effectiveness acceptability
curve, which shows the chances that a treatment is cost-effective relative
to another for different levels of willingness to pay. The cost-effectiveness
acceptability curve shows the probability (on the y-axis) that PGx may be
cost-effective compared with N-PGx for a range (on the x-axis) of maximum monetary values that a decision-maker might be willing to pay per
QALY. According to Figure 8.7A *for the classical model (assuming linear shar-
ing between additional cost and effectiveness units without upper limit across ΔE),*
the probability of PGx being cost-effective at €60,000 per QALY, is more
than 80%.

In the case of the new model, assuming $\Delta E_{max} = 0.027$ or 0.01 and

$$\lambda = \frac{1 - e^{(-5\Delta E^3)}}{\Delta E} 310$$

where $\Delta C_{max} = $€310.0, different acceptability curves are determined.
The results of the probabilistic analysis are shown in Figure 8.7B. It must
be noted that the scenarios with $\Delta E_{max} = 0.027$ are too "ambitious" and
"demanding" because they require that PGx *must* have a difference of
0.027 QALYs against N-PGx to invest all the additional €310, and thus

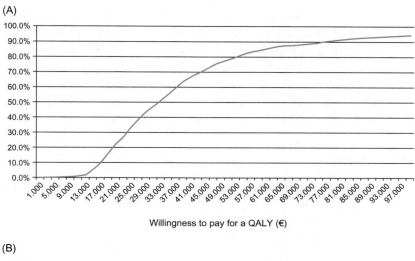

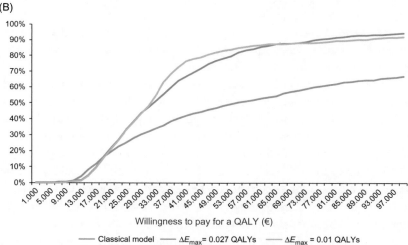

Figure 8.7 *Cost-effectiveness acceptability curve for PGx group versus the N-PGx control group using the classical model (A) and the new model (B).* $\Delta E_{\max}$: maximum expected difference in effectiveness between the N-PGx and PGx groups, which could be additionally reimbursed given that the budget is 0.01 and 0.027 QALYs in the two scenarios, respectively. λ: $((1 - e^{(-5\Delta E^{3})})/\Delta E)310$ for the new models.

the probability of PGx being cost-effective is lower across willingness to pay for a QALY against the classical model that is more conservative. On the contrary, the scenario with $\Delta E_{\max} = 0.01$ represents a society with low expectations against the innovation, because the amount of €310 will be invested even if the new technology (PGx) has only an incremental difference of 0.01 QALYs against N-PGx (closer to classical model). From this example, it must be clear that the probability of a new technology being cost-effective is a function of the shape of λ across willingness to pay (namely the parameters a and b) and the maximum amount a budget holder is willing to pay to capture the maximum expected additional effectiveness (namely is a function of ΔE_{max} and ΔC_{max}).

IMPLICATIONS

Today, the cost of innovation in health care is expected to be covered by the third-party payers and usually leads to uncontrolled growth in health care expenditure. In view of the scarcity of resources, economic evaluation provides a criterion for the final decision concerning the adoption of certain new interventions but has certain inconsistencies and drawbacks. In this analysis, a new methodological approach was proposed taking into account the budget constraint, the effectiveness of a new intervention, and the willingness to pay in a flexible way.

In this new cost-effectiveness analysis model, it was assumed that the budget is exogenous and has been set by the budget holders. It must be noted that the health care budgets frequently depend on historical, political, and social criteria without taking into account the economic models and their assumptions (Schwappach, 2002). Furthermore, uncertainty in the determination of costs and outcomes (Ramsey et al., 2005), lack of data or knowledge for preferences (Weyler and Gandjour, 2011), lack of training of policymakers (Veney et al., 1997), established status quo, and other policy relevant issues set restrictions in a rational decision-making process for the health maximization problem. Thus, the quantitative determination of constraints and the application of prominent instruments such as mathematical programming (Flessa, 2000) or other methodological approaches (Sendi and Briggs, 2001) seem difficult to be implemented practically for the optimum allocation of the health care budgets.

In this model, a far less ambiguous approach was adopted to estimate the equilibrium that is driven by the *de facto* budget availability. The model allows the determination of λ (via the budget constraint and the S-curve for λ) to vary across ΔE. This argument of varying determination of λ is in accordance with related literature (Bridges et al., 2010; Zhao et al., 2011). There are some notable differences between the proposed new and classical models. In the proposed new model, the maximum incremental effectiveness that could be additionally reimbursed by payers is predetermined, taking into account the budget constraints. The model presented here describes, in a dynamic manner, the link between the willingness to pay threshold toward the budget limit, which can be particularly helpful in developing and resource-limited countries, where there are often significant budget limitations. In addition, the equilibrium between λ, ΔE, and B_t is in accordance with a variable return on scale assumption of λ, taking into consideration the size of ΔE and, partly, the degree of innovation for a new intervention. Moreover, the classical approach (fixed λ) has been incorporated as a particular case, given that we could find appropriate coefficients in the Weibull equation that produce almost a straight line segment in the cost–effectiveness plane.

The methodology, proposed in this chapter, shares the spirit of methods used in microeconomic theory. The model flexibly demonstrates how willingness to pay changes as the budget remains fixed and allows ΔE to vary, how the budget must be expanded to afford a new technology given a specific willingness to pay path, fixed or not, or a combination of these. This approach has some degree of generalization because it could possibly be considered as a "standard intervention" in the bundle of interventions that have already launched in the market and have been reimbursed by the payers. Also, it should be noted that this model does not take into account differences in the incidence of pharmacogenomic biomarkers, which may vary significantly among different populations and ethnic groups (Georgitsi et al., 2011).

The analysis focuses mainly on the presentation of results deterministically and gives less attention to estimating the equilibrium in a stochastic manner. In such a case, the solution of the model could be estimated with some degree of uncertainty depending on the assumption made for the parameter of interest or the data availability. Inevitably, the implementation of the solution for practical purposes becomes much more complex when taking into account the uncertainty of all variables.

CONCLUSIONS

The model presented in this chapter provides a new conceptual methodology for the betterment of our understanding in decision-making process, as well as a practical tool for educational purposes. The simplification of the model's assumptions and the calibration of the

146 Economic Evaluation in Genomic Medicine

mathematical equations to reflect more accurately the real-life setting should very well be the scope of future research.

Overall, this new model would be particularly useful to perform economic evaluation studies of novel genomic medicine interventions in different countries and health care systems, taking into account health care budget differences, so that resource allocation in these countries could be optimized.

REFERENCES

Barton, G.R., Briggs, A.H., Fenwick, E.A., 2008. Optimal cost-effectiveness decisions: the role of the cost-effectiveness acceptability curve (CEAC), the cost-effectiveness acceptability frontier (CEAF), and the expected value of perfection information (EVPI). Value Health 11 (5), 886−897.

Birch, S., Gafni, A., 2006. The biggest bang for the buck or bigger bucks for the bang: the fallacy of the cost-effectiveness threshold. J. Health Serv. Res. Policy 11 (1), 46−51.

Bridges, J.F., Onukwugha, E., Mullins, C.D., 2010. Healthcare rationing by proxy: cost-effectiveness analysis and the misuse of the $50,000 threshold in the US. Pharmacoeconomics 28 (3), 175−184.

Cartwright, W.S., 1999. Methods for the economic evaluation of health care programmes, second edition. By Michael F. Drummond, Bernie O'Brien, Greg L. Stoddart, George W. Torrance. Oxford: Oxford University Press, 1997. J. Ment. Health Policy Econ. 2 (1), 43.

Donaldson, C., Birch, S., Gafni, A., 2002b. The distribution problem in economic evaluation: income and the valuation of costs and consequences of health care programmes. Health Econ. 11 (1), 55−70.

Donaldson, C., Currie, G., Mitton, C., 2002a. Cost effectiveness analysis in health care: contraindications. BMJ 325 (7369), 891−894.

Eckermann, S., Briggs, A., Willan, A.R., 2008. Health technology assessment in the cost-disutility plane. Med. Decis. Making 28 (2), 172−181.

Flessa, S., 2000. Where efficiency saves lives: a linear programme for the optimal allocation of health care resources in developing countries. Health Care Manage. Sci. 3 (3), 249−267.

Gafni, A., 1998. Willingness to pay. What's in a name? Pharmacoeconomics 14 (5), 465−470.

Gafni, A., Birch, S., 2006. Incremental cost-effectiveness ratios (ICERs): the silence of the lambda. Soc. Sci. Med. 62 (9), 2091−2100.

Georgitsi, M., Viennas, E., Gkantouna, V., Christodoulopoulou, E., Zagoriti, Z., Tafrali, C., et al., 2011. Population-specific documentation of pharmacogenomic markers and their allelic frequencies in FINDbase. Pharmacogenomics 12 (1), 49−58.

Gyrd-Hansen, D., 2005. Willingness to pay for a QALY: theoretical and methodological issues. Pharmacoeconomics 23 (5), 423−432.

Kvamme, M.K., Gyrd-Hansen, D., Olsen, J.A., Kristiansen, I.S., 2010. Increasing marginal utility of small increases in life-expectancy? Results from a population survey. J. Health Econ. 29 (4), 541−548.

McCabe, C., Claxton, K., Culyer, A.J., 2008. The NICE cost-effectiveness threshold: what it is and what that means. Pharmacoeconomics 26 (9), 733−744.

Nelson, A.L., Cohen, J.T., Greenberg, D., Kent, D.M., 2009. Much cheaper, almost as good: decrementally cost-effective medical innovation. Ann. Intern. Med. 151 (9), 662–667.

O'Brien, B.J., Briggs, A.H., 2002. Analysis of uncertainty in health care cost-effectiveness studies: an introduction to statistical issues and methods. Stat. Methods Med. Res. 11 (6), 455–468.

O'Brien, B.J., Gertsen, K., Willan, A.R., Faulkner, L.A., 2002. Is there a kink in consumers' threshold value for cost-effectiveness in health care? Health Econ. 11 (2), 175–180.

Ramsey, S., Willke, R., Briggs, A., et al., 2005. Good research practices for cost-effectiveness analysis alongside clinical trials: the ISPOR RCT-CEA Task Force report. Value Health 8 (5), 521–533.

Schwappach, D.L., 2002. Resource allocation, social values and the QALY: a review of the debate and empirical evidence. Health Expect. 5 (3), 210–222.

Sendi, P., Gafni, A., Birch, S., 2002. Opportunity costs and uncertainty in the economic evaluation of health care interventions. Health Econ. 11 (1), 23–31.

Sendi, P.P., Briggs, A.H., 2001. Affordability and cost-effectiveness: decision-making on the cost-effectiveness plane. Health Econ. 10 (7), 675–680.

Stinnett, A.A., Paltiel, A.D., 1996. Mathematical programming for the efficient allocation of health care resources. J. Health Econ. 15 (5), 641–653.

Towse, A., 2009. Should NICE's threshold range for cost per QALY be raised? Yes. BMJ 338, b181.

Veney, J.E., Williams, P., Hatzell, T., 1997. In search of failure: guidelines for ministries of health. J. Health Popul. Dev. Ctries. 1 (1), 1–15.

Wailoo, A., Tsuchiya, A., McCabe, C., 2009. Weighting must wait: incorporating equity concerns into cost-effectiveness analysis may take longer than expected. Pharmacoeconomics 27 (12), 983–989.

Weinstein, M., Zeckhauser, R., 1973. Critical ratios and efficient allocation. J. Public Econ. 2, 2147–2157.

Weyler, E.J., Gandjour, A., 2011. Empirical validation of patient versus population preferences in calculating QALYs. Health Serv. Res.

Whitehead, S.J., Ali, S., 2010. Health outcomes in economic evaluation: the QALY and utilities. Br. Med. Bull. 96, 5–21.

Yin, X., Goudriaan, J., Lantinga, E.A., Vos, J., Spiertz, H.J., 2003. A flexible sigmoid function of determinate growth. Ann. Bot. 91 (3), 361–371.

Zhao, F.L., Yue, M., Yang, H., Wang, T., Wu, J.H., Li, S.C., 2011. Willingness to pay per quality-adjusted life year: is one threshold enough for decision-making? results from a study in patients with chronic prostatitis. Med. Care 49 (3), 267–272.

CHAPTER 9

Conclusions and Future Perspectives

INTRODUCTION

The previous chapters outlined the various technical aspects and selected applications of economic evaluation in genomic medicine. If applied correctly, Cost-effectiveness analysis (CEA) and Cost—utility analysis (CUA) are promising technical analyses, but they are also characterized by certain drawbacks. Some of these can be resolved if handled appropriately; others involve political issues and conflicts of interest characteristic of the health care sector in general. There are also challenges associated with the nature of the subject itself.

Lack of Education

In many cases, health care decision-makers cannot understand the methodological aspects of economic evaluation if they are not familiar with them, which may lead to a lack of trust in the results. For example, they are unable to understand basic concepts, such as the quality-adjusted life-year (QALY), or the fact that a cost-effective treatment does not necessarily mean savings in resources and that it could be just the "socially acceptable option"—much less the concept of the "acceptability curve." They may also conclude that the results are not relevant because the analyses were not done from the perspective of "their system."

This problem is soluble in that policymakers can familiarize themselves with the principles of economic analysis and can utilize experts conversant with this scientific field and "translate" research findings into a language that can be understood by the general public.

This Field Is Highly Technical but Not Very "Objective"

In recent years, CEA has become a highly technical field with many links to mathematics and statistics, which means that nonexperts find it difficult to understand, evaluate, or contribute to it. This problem is further aggravated by the fact that conventional medical curricula do not include such subjects, and doctors first come in contact with this field only after completing their formal studies. Obviously, in this case valid doubts are raised regarding the reliability of the analyses because those responsible for applying the method's conclusions are unable to validate them. Such an overly technical analysis also has an additional problem: it allows the health economists to modify a statistical approach and to be led to a completely different result depending on the assumptions implemented for their model. Lack of evidence for interventions, which is a particular challenge in genomic medicine, increases the reliance on assumptions that, if incorrect, can lead to misleading results. Application of sensitivity analysis can only partially account for this problem.

There Is No Standard Methodology for Study Performance/ Presentation

Some have also claimed that economic evaluation suffers from a lack of standardized methodology in performing the published studies. Even among the pioneers of the field, one finds examples with differing results depending on the assumptions made, a fact indicating that it is difficult to compare economic studies, particularly if the publication does not adequately present the methods and assumptions to allow independent analysis. The standard literature exacerbates this problem with limitations on word count, tables, and figures, although the ability to publish supplemental materials online can ameliorate this issue. Clinical effectiveness is connected to physical reality and can be perceived easily, whereas the standardized quantification of abstract economic quantities (e.g., loss of productivity due to death) is much more difficult. We should mention here that criticism of this issue has been productive and efforts are being made by the scientific community to standardize the methodology. In addition, many reputable journals specify important conditions with regard to paper presentation to ensure that they are clear, ethical, reproducible, and comprehensible. If someone were to follow the evolution of the subject over the past 10–15 years, they would conclude that simplistic and occasionally misleading analyses have been replaced by more papers that are sophisticated, more definite, and objective, especially in the leading journals.

Of course, one should remember that the nature of the subject itself is complex and, in theory, attempts to answer a much more ambitious question than a clinical study because it combines cost and effectiveness data; furthermore, it attempts to standardize, in a unit model, conclusions of uncertainty involving the entire society.

There Is Bias Because of Funding

This is a potential concern because companies funding such studies are for-profit enterprises and seek to depict their products as useful for their "clientele" (the medical community or the insurance funds). The same argument could also hold for clinical or other trials and is the subject of increased scrutiny. In particular, creation of clinical guidelines by professional societies has been noted to be dependent on nonobjective methods that are subject to bias. The recent publication from the Institute of Medicine (Clinical Guidelines We Can Trust; Graham et al., 2011) outlines an approach to reduce the risk of bias in the guidelines. Similar recommendations could be created for economic analysis to address these concerns (Berger and Olson, 2013). Journals have increased the requirements for disclosure of potential conflict of interest and funding sources to enhance transparency and reduce the publication of biased analyses. Also, as the subject "matures," the companies themselves understand that an impartial analysis is more beneficial in the long-term for their products' image than any other approach. In fact, some companies are using economic analyses to set prices based on cost-effectiveness thresholds. Of course, the authors' commitment to the study sponsors is also a limitation, because they are asked to sign confidentiality agreements and waive the right to publish their findings without authorization. This can result in unfavorable results not being submitted for publication, although journals have also been shown to be less likely to publish studies with negative results (Knight, 2003).

It Is Based on Multiple Assumptions

In many cases, CEA is "accused" of being dependent on too many assumptions for its models; therefore, its results cannot be used in real life by clinical scientists or policymakers. Economic evaluation often collects data from many sources to reach a conclusion. The unsuspecting reader might believe that the results of the analysis are a "universal truth" even when the model is beset with uncertainties. For example, in many cases we are interested not only in comparing the effectiveness of two interventions in a clinical study but also in the long-term estimation of the incremental cost-effectiveness ratio (ICER) for the same patients. In such cases, it is necessary to combine data from the clinical study with observational or registry data, in which case the degree of uncertainty increases. One solution to this problem is to perform a CEA based on the clinical trial patient data and then perform a supplemental sensitivity analysis to calculate the long-term results. Another good and reliable solution is to explicitly declare the assumptions of statistical models and the way that these affect the results. It is always important to describe what programs were used to perform the analysis, such as Excel or other programs.

It Uses a Different Perspective than the One Commonly Used by Clinical Scientists or Policymakers

The estimation of the cost-effectiveness ratio is a primary goal of economic evaluation; however, what politicians care about in most cases is "reduced expenditures," not "increased expenditure based on social criteria." The politician's primary goal is not to exceed the available budget, whereas the clinical scientist cares about prescribing the treatment that will maximize the patient's outcomes, including survival. The health economist stands between the two, trying to mediate the two viewpoints.

Reproduction of Authority by the Medical Community

CEA is an "innocent" technical field of analysis, but it can actually and indirectly set some limits on a doctor's "authority" over patients and society; therefore, its widespread application in practice has met with resistance. For the first time, we stand at the brink of an important development: a scientist outside the medical community (an economist) "criticizes," through their results, the practice of clinical scientists without having the clinical background to correctly understand the consequences of the theories they propose. In the United States, this was seen in the enabling legislation that created and funded the Patient-Centered Outcomes Research Institute (PCORI), which specifically prohibited funding for cost-effectiveness studies. Although this was a political decision, it reflected an underlying concern that cost-effectiveness would interfere with the practice of medicine at the discretion of physicians and would interfere with the development and implementation of new drugs and technologies. We believe that productive use of economic evaluation is achieved when it is performed and evaluated with the help of scientists from many different specialties, so that the study results have greater validity and greater probability of being accepted by a large number of health professionals.

It Is not "Ethical"

Economic evaluation seeks to maximize social welfare through statistical approaches, whereas clinical scientists seek to maximize the welfare of their patient (personalized medicine). In this sense, the clinical scientist's approach can never fully coincide with the economist's because it has been said that it is "unethical" to prescribe a treatment that we know not to be the most effective one based on pharmacoeconomic criteria. To this end, there are several ethical issues that are complex and therefore beyond the scope of this book.

It Increases Costs

It has also been said that economic evaluation is mostly used by companies to convince insurance funds to reimburse their expensive treatments without specifically indicating which competing treatment should be rejected from the reimbursement system to meet the relevant budget goals. Studies proposing immediate discontinuation of reimbursement of specific treatments that are not cost-effective are rare or nonexistent. The analysis usually proceeds in only one direction: the introduction of even more treatments. This puts tremendous pressure on insurance funds, which collapse under the burden of the economic obligations created by the expectations of patient, providers, and policymakers.

Despite the negative statements regarding the disadvantages of economic evaluation, it should be clear from this book that there is potential for its utilization in the current political climate. Preparing an economic evaluation is now a prerequisite for a medication to be reimbursed in most European countries, which is a significant sign of the acceptance of this approach. In several countries, the first steps have already been taken and we are quickly proceeding in the direction of full implementation. This development creates a basis that will help all shareholders in the health care sector take into account the cost of the treatments they offer, along with their benefit.

We believe that for new treatments it is very important to prepare economic evaluations, both to estimate the cost-effectiveness ratio and to determine their burden on the health system. This is not as important for already existing health technologies, because these are established in clinical practice and any studies performed are not expected to lead to significant changes in practice.

For a new treatment, things are quite different. In cases of "marginal" innovation from a new technology, economic evaluation could be used as a means of pressure on the industry to reduce prices; however, in the case of actually innovative products, it would be a means to reward the innovators. An important step would be the founding of an academic, autonomous agency that would evaluate health technologies and would provide its conclusions to the political authorities. In this case, a careful reading of economic evaluations could potentially form a natural nucleus from which to begin negotiations between state and industry.

CONCLUSIONS

Economic evaluations are of utmost importance for genomic medicine to demonstrate not only the cost–effectiveness of a certain intervention that directly impacts on the national health care expenditure but also, most importantly, its potential to improve the quality of life of the patients. This is an emerging and highly promising field (Snyder et al., 2014) considering the very few economic evaluation studies in pharmacogenomics and genomic medicine.

Furthermore, there is an urgent need at the moment to perform economic evaluation studies in various developing countries to demonstrate the cost–effectiveness of the individualization of certain therapeutic interventions that may not necessarily be cost–effective in developed countries. For example, individualization of warfarin treatment has been shown to be cost–effective for elderly Croatian patients (Mitropoulou et al., 2015), suggesting that, contrary to other developed countries, pharmacogenomics-guided warfarin treatment not only represents a cost–effective therapy option for the management of elderly patients with atrial fibrillation in Croatia but also may be the case for the same and other anticoagulation treatment modalities in neighboring countries.

This fact dictates such economic evaluation studies to be replicated in every health care system. To this end, a generic economic evaluation model would be extremely useful to assess cost–effectiveness of individualized treatment modalities. Such a tool could be used to perform a cost–effectiveness threshold analysis across a variety of input conditions, such as drugs, costs, and biomarkers, and would enable entities with no specific expertise in economic analysis but with access to the necessary parameters and cost items (see also Chapter 7) to reach a rapid decision regarding whether a certain genome-based therapeutic intervention is cost-effective in various settings/countries.

Considering the large economic evaluation evidence gaps in genomic medicine, there is an urgent need to adequately address them to measure the value of current and emerging genomic technologies, which would allow appropriate value-based decisions about adoption and investment in genomic medicine interventions. In essence, realizing the huge potential of genomic medicine to benefit society, it may be of equal importance to engage not only in innovative scientific discovery in genomics research alone but also, most importantly, in scientific research to produce break-throughs in the field of economic evaluation of genomic medicine, demonstrating its value to those who will support and fund its integration into mainstream medical practice.

REFERENCES

Berger, A.C., Olson, S., 2013. The Economics of Genomic Medicine: Workshop Summary. Institute of Medicine of the National Academies, National Academies Press, Washington, DC.

Graham, R., Mancher, M., Miller-Wolman, D., Greenfield, S., Steinberg, E., 2011. Clinical Practice Guidelines We Can Trust. Institute of Medicine of the National Academies, National Academies Press, Washington, DC.

Knight, J., 2003. Negative results: null and void. Nature 422, 554–555.

Mitropoulou, C., Fragoulakis, V., Bozina, N., Vozikis, A., Supe, S., Bozina, T., et al., 2015. Economic evaluation for pharmacogenomic-guided warfarin treatment for elderly Croatian patients with atrial fibrillation. Pharmacogenomics.

Snyder, S.R., Mitropoulou, C., Patrinos, G.P., Williams, M.S., 2014. Economic evaluation of pharmacogenomics: a value-based approach to pragmatic decision-making in the face of complexity. Public Health Genomics 17, 256–264.

INDEX

Note: Page numbers followed by "*f*" and "*t*" refer to figures and tables, respectively.

Printed in the United States
By Bookmasters